How
Animals
Live
edited by

Peter
Hutchinson

*a series
of volumes
describing
the behaviour
and ecology
of the animal
kingdom*

VOLUME 1

Maurice Burton

How Mammals Live

ELSEVIER  PHAIDON

Credits
to Photographers

Artists and photographers are listed alphabetically with their agent's initials, where applicable, abbreviated as follows: (AFA) Associated Freelance Artists Ltd; (Ardea) Ardea Photographics; (NSP) Natural Science Photos; (Res) Bruce Coleman Ltd.

H. Albrecht (Res) 69 140
M. E. Bacchus (Res) 22
D. Bartlett (Res) 42 114
P. A. Bowman (NSP) 92 113
G. T. Bowra (AFA) 153
J. Brennan (NSP) 51
J. Burton (Res) 2 16 18 33 79 99 105 141 149(top) 151
P. J. K. Burton (Res) 116
B. Campbell (Res) 25 101 148
B. Coleman (Res) 143
N. Duplaix 131
Elsevier 8–9 20(2) 28 31 32 48 50 81 110 118 133 136 138 144
Dr. J. B. Free 97
C. Frith (NSP) 63
H. Gauthier-Pilters 39
S. Gillsater (Res) 128
C. M. Hladik 15 21 61 139 150
E. Hosking 38 54 88(2) 95 154
P. Jackson (Res) 41 58 65
M. Kawai 76
R. Kinne (Res) 57 73 120 130

G. Kinns (AFA) 17 23 40 62 100 104 155
Dr. H. Klingel 56
D. B. Lewis (NSP) 84
A. Leutscher (NSP) 112
J. Markham 67(top)
Dr. R. D. Martin 82, 135
T. McHugh (Res) 93
R. Moreton (AFA) 80
Dr. P. A. Morris 30 64 91 149(bottom)
N. Myers (Res) 29 37 83
E. C. Neal (Res) 108
D. Ovenden 10 43 59 85 111 117 137
D. M. Patterson (Res) 55
G. Pizzey (Res) 119 126 127
Polfoto (AFA) 71
Dr. D. Pye 121
M. Quarishy (Res) 49 66 147 156
D. Robinson (Res) 12 27 109
A. Root (Res) 35
F. Rotgans 98 102–103(top)
L. L. Rue (Res) 53 67(bottom) 87 125
Professor D. R. Schenkel 89
A. J. Sutcliffe (NSP) 26
B. Thomas (Res) 129
S. Trevor (Res) 47 107
A. van den Nieuwenhuizen 6 44 75 77 123
Drs. A. A. M. van der Heyden 36 45 60 102(bottom)
R. W. Vaughan 19 72
P. H. Ward (NSP) 146

Elsevier-Phaidon
An imprint of Phaidon Press Ltd.
Published in the United States by E.P. Dutton & Co. Inc.,
201, Park Avenue South, New York, N.Y. 10003

First published 1975
Reprinted 1977
Planned and produced by
Elsevier International Projects Ltd, Oxford
© 1975 Elsevier Publishing Projects SA, Lausanne.

ISBN 0 7290 0021 4

Filmset by Keyspools Limited, Golborne, Lancashire
Printed and bound by Brepols - Turnhout - Belgium

Contents

Introducing the Mammals

There are between 4,000 and 5,000 species of mammals living today, of which two-thirds are rodents. This is a small total compared with other groups – 6,000 reptiles, nearly 9,000 birds, over 20,000 fishes and nearly 1,000,000 insects. Yet, because mammals are often of large size, they tend to be the more familiar animals. They are also more easily understood because, unlike other groups that capture the imagination, such as the dinosaurs, mammals have a form, structure and way of life that most closely approximates to our own. This is hardly surprising since the human species itself is included in the class Mammalia.

The mammals are vertebrates and are classified in the group of animals called Vertebrata. Unlike the vast majority of animals, the invertebrates, the body of all members of this group is supported by a backbone of elements called vertebrae that are usually composed of bone, but sometimes of cartilage. Other vertebrates are the fish, amphibians, reptiles and birds, but the mammals are quite distinct from these groups because they have evolved a complex of characteristics that is unique. They are warm-blooded, usually with a body covering of hair, and the young are usually born alive and nourished by milk secreted from glands in the female body. These glands are normally on the under or ventral side of the body but exceptionally may be nearly in the arm-pits or high up on the sides of the body, as in the coypu, the large South American rodent, which nurses its young as it swims.

The Origin of Mammals. Surprisingly, such varied mammals as mice, elephants, whales and men, have evolved from the same type of animal, small and shrew-like, that lived almost 200 million years ago. These animals (called *Morganucodon* by some zoologists and *Erythrotherium* by others) had some characteristics that are found in reptiles and others that are typically mammalian. We know from studying the fossil remains of mammals that the Mammalia evolved from early reptiles and, although

The Common bush-baby, *Galago senegalensis*, unlike its close relatives the lorises, angwantibo and pottos, is an agile tree dweller which must explore its habitat before it can jump from branch to branch with safety. Its tree dwelling habits are reflected by its large eyes and grasping hands.

these two groups are distinct today, at some period during the history of the vertebrates, intermediate groups must have lived. *Morganucodon* is one of these intermediate forms.

The mammals living today are classified into three groups or orders; the Monotremata or monotremes; the Marsupialia or marsupials; and the Placentalia or placentals. The Monotremata is contained in the subclass Prototheria, and Marsupialia and Placentalia in the subclass Theria. To understand why this is so, we must trace the evolutionary history of the mammals.

The dominant land vertebrates some 250 million years ago were the reptiles. Of the tremendous diversity of forms then alive, only the crocodiles, turtles, tortoises, snakes and lizards remain today. These animals, like their ancestors, are limited in three important ways; they reproduce by laying eggs, they have little control over their body temperature, and they have small brains and are therefore incapable of very complex behaviour patterns or of adapting their behaviour to changed circumstances.

It is very likely that one of the reptilian features of *Morganucodon* was that it laid eggs. When the first monotremes evolved from *Morganucodon*, they too laid eggs. We know this because the last surviving monotremes, the duckbill or platypus and the several species of Spiny anteater, lay eggs even today. The monotremes exhibit many other reptilian features, as in bones of the shoulder girdle and the shape of the limb bones, the cartilaginous cup in the eye, found in reptiles but not in mammals, and the shape of the cochlear bone of the inner ear. Another reptilian feature is the possession of a cloaca through which the genital, urinary and faecal products pass.

Within a hundred million years of their origins, a new method of reproduction evolved in mammalian stock. Instead of growing inside an egg, the young were born at an early stage of development and thereafter were carried attached to the maternal teats, usually within a pouch on the mother's abdomen. This early birth was necessary because there was either no placenta or no well-developed placenta through which the embryo could be

7

PRESENT DAY

PLEISTOCENE

PLIOCENE

Glyptodon

MIOCENE

edentates lagomorphs rodents primates bats

OLIGOCENE

EOCENE

PALEOCENE

CRETACEOUS

The probable evolution of the mammals. The earliest known fossils of mammals date back to early Mesozoic times and are about 200 million years old. These fragmentary fossils are mainly of skulls and lower jaws of shrew-like animals. From them have descended all the placental mammals, the probable evolution of which is depicted in this chart. It can be seen that most of the major mammalian groups have been in existence since the Oligocene, 40 million years ago. The width of each orange area is drawn in proportion to the numbers of families that lived at a given time. In general, the numbers of mammalian families has declined since the ice ages of the Pleistocene.

nourished from the mother's body. This method of reproduction was a distinct advance over the monotreme method, for it afforded more protection for the young at the most crucial early days of their lives, and it is still employed by members of the second order of mammals, the Marsupialia. Like the monotremes, the marsupials also have a few reptilian features, but the fact that they are warm-blooded and have a covering of hair, emphasises that there can be no doubt that they are mammals.

The third mammalian order, the Placentalia, contains the vast majority of mammals living today. In all, the young are retained inside the mother for a long period prior to birth, during which the

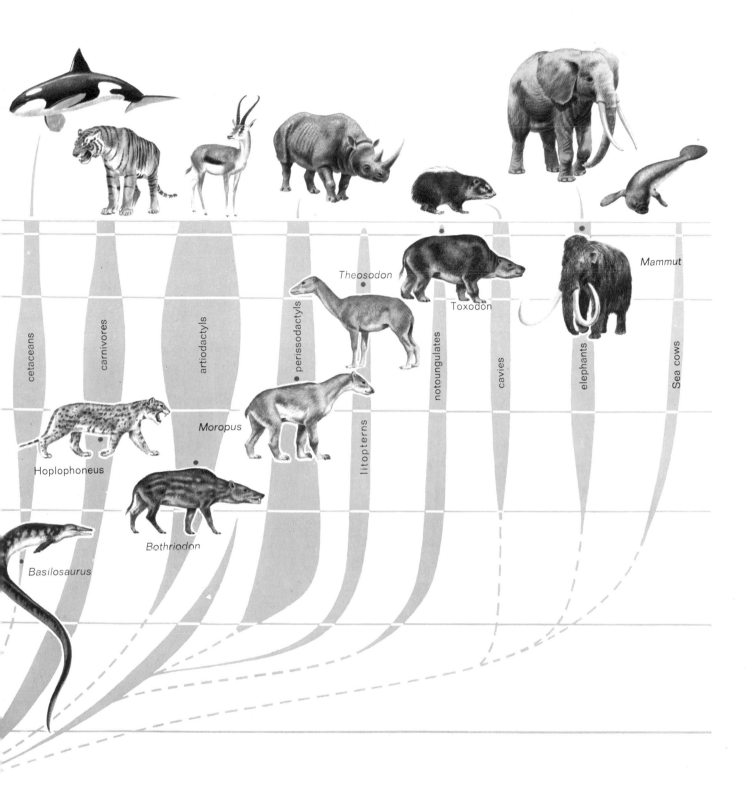

cetaceans

carnivores

artiodactyls

perissodactyls

Theosodon

litopterns

notoungulates

Toxodon

cavies

elephants

Mammut

Sea cows

Moropus

Hoplophoneus

Bothriodon

Basilosaurus

developing embryo derives all its nourishment from the mother. It was the evolution of this extremely successful method of reproduction that enabled the placentals to colonize a great variety of habitats: bats in the air; whales and dolphins in the sea; otters and beavers in rivers; and mice and men on land.

Characteristic Features of Placental Mammals. The most important diagnostic feature of the placentals is, as mentioned above, their method of reproduction. This will be discussed in more detail in Chapter 6. Other features include the presence of hair that acts as an insulating cover in most placentals, three bones, the auditory ossicles, that

9

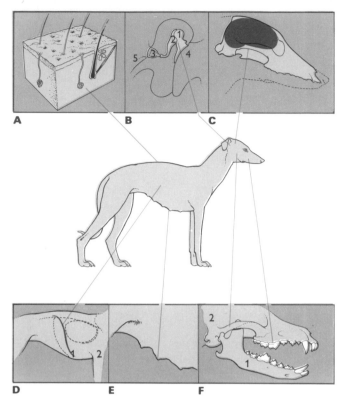

The mammals are defined by a number of anatomical features that are not found in other animals. The more important ones are: (A) the presence of hair; (B) three bones in the inner ear, the malleus (1), incus (2) and stapes (3), which transmit vibrations from the ear drum (4) to the oval window (5); (C) a large brain; (D) a diaphragm (1) which assists ventilation of the lungs (2); (E) mammary glands in which milk is produced; and (F), a lower jaw composed of a single bone called the dentary (1), which articulates with the squamosal (2), and teeth which are modified to perform a variety of functions.

transmit vibrations across the inner ear, and a lower jaw that is composed of a single bone called the dentary. Another feature of placentals is the large size of the brain, important as the evolution of a large brain led to intellectual versatility that made more varied behaviour possible. In particular it led to the development of a sense of curiosity.

At least since Darwin's day it has been realised that there are three basic requirements in the life of an animal; the search for food, the need to reproduce and to escape from enemies, in that order of importance. Only a quarter of a century ago was it appreciated that there existed a fourth requirement, which was broadly named the sense of curiosity. In the chapters that follow, we will see how important is a sense of curiosity to mammals.

The Classification of Mammals. To a large extent the classification of mammals is based on the characteristics of their teeth, although these are absent in forms such as monotremes and some of the anteaters. Mammals have different kinds of teeth that perform different functions: spatulate or chisel-like incisors for biting, long pointed canines for stabbing, and flat cheek teeth (molars and premolars) with cusps for chewing. The incisors are in the front of the jaw, the canines at the 'corners' and the cheek teeth along the sides. The full complement of teeth is found in the Insectivora, Tupaioidea, Chiroptera and Primates. The rodents lack canines but have specialized incisors, as do rabbits and hares. Carnivores typically have modified molar and premolar teeth, the carnassials, for slicing flesh, and herbivores, have grinding molars, for dealing with tough vegetation.

The most remarkable dentition is that of elephants and Sea cows. Apart from its tusks, the elephant has seven molars in both lower and upper jaws of each side during the course of a lifetime. The first of the seven is a milk tooth and is soon lost. The other six come successively forward in the jaw, on a 'conveyor belt' system, only one being in use at a time. When each is worn down it is cast, and the new tooth behind replaces it. When the last molars are worn down, the elephant can no longer feed and is therefore doomed to die.

Although the classification of the mammals is based on the teeth, adjustments need to be made for the exceptions and paradoxes. For example, the hyrax looks like a rodent and has a single pair of incisors in the upper jaw, like the guinea pig it resembles, but it has four incisors in the lower jaw. It also has a gap between the incisors and cheek teeth, but the latter are patterned, somewhat surprisingly, as in the rhinoceros. Thus, although it could easily be mistaken for a rodent, the hyrax is placed in an order on its own, near the rhinoceros, and also near the elephants, which it resembles in other features of its anatomy.

Another exception to the rule is seen in the order Carnivora that contains dogs and cats as well as other flesh-eaters. The exception is the Giant panda which eats largely bamboo shoots. In it, the dentition is mainly of the carnivorous type but with the cheek teeth either lacking the shearing cups or having them only feebly developed. The molars have almost flattened crowns, for grinding tough, fibrous plant foods.

So, while the experienced zoologist can make a good guess as to which order a mammal belongs by looking at its teeth, he does so knowing not only the general rules, but also where the exceptions lie.

Classification of the Class Mammalia

SUBCLASS	ORDER	EXAMPLES
PROTOTHERIA	MONOTREMATA	duckbill or platypus, Spiny anteater
THERIA	MARSUPIALIA	kangaroo, wallaby, wombat, koala
	INSECTIVORA	shrew, mole, hedgehog
	TUPAIOIDEA	tree-shrew
	DERMOPTERA	Flying lemur
	CHIROPTERA	bats
	PRIMATES	lemur, monkey, ape, man
	EDENTATA	anteater, armadillo, sloth
	PHOLIDOTA	pangolin
	LAGOMORPHA	pika, rabbit, hare
	RODENTIA	rats, mice, squirrel, hamster
	CETACEA	whale, dolphin, porpoise
	CARNIVORA	dog, wolf, fox, jackal, cat, lion, tiger, leopard, lynx, weasel, stoat, badger, otter, skunk, panda, raccoon, aardwolf, hyaena, bear, mongoose, civet, genet
	PINNIPEDIA	seal, Sea lion, walrus
	TUBULIDENTATA	aardvark
	PROBOSCIDEA	elephant
	HYRACOIDEA	hyrax
	SIRENIA	dugong, manatee
	PERISSODACTYLA	horse, ass, zebra, tapir, rhinoceros
	ARTIODACTYLA	pig, hippopotamus, camel, sheep, goat, antelope, deer, giraffe

Extinct mammalian groups not included

Exploratory Behaviour and the
Sense of Curiosity

Several years ago, in my garden in the south of England, a score of Muscovy ducks were wandering about looking for food or merely wandering about. In one corner of the garden was a large oblong cage in which lived a tame Grey squirrel. The squirrel started to play on the ground, tumbling and somersaulting. The nearest duck was not less than 20 ft (6 m) from the squirrel, the farthest probably 40 ft (12 m). In no time at all from the moment the squirrel began its antics, the ducks, as if with one accord, started to waddle direct to the squirrel's cage. There they ranged themselves around two sides of the cage, each duck with its beak – and therefore its eyes – directed towards the performing squirrel. When the play stopped and the squirrel retired to a high perch, the ducks slowly moved away and went back to what they were doing before the interruption occurred.

This is merely an example of a group of animals, drawn by sheer curiosity, that went out of their way to watch a totally dissimilar animal whose behaviour was out of the ordinary. It is not only an example of pure curiosity, it is an example of a situation that could, under other circumstances, be exploited. It helps the understanding of reported instances in which carnivores have used such situations to effect a kill, by resorting to behaviour known as 'charming'.

There are many records of the Red fox and the stoat indulging in this behaviour. A typical example is as follows. A man is riding through a wood when he suddenly reins in his horse, his attention caught by a fox in a clearing which seems to have taken leave of its senses. The fox is prancing around in small circles, then tumbling and somersaulting, much in the way the Grey squirrel was behaving when the Muscovy ducks gathered round it. In a circle around the fox is a number of birds of various sizes and perhaps a rabbit or two. As the man watches the birds and rabbits edge towards the fox, staring at it as if fascinated. Then the fox suddenly ceases

its clowning and pounces on one of the spectators, seizing it.

It has always been a matter of opinion how this should be interpreted. Was the fox merely playing, as foxes in captivity frequently do, with no evil intent? Or does a fox use this as a deliberate stratagem to obtain an easy meal? Since a fox typically hunts at night the more likely explanation is that the fox is playing but, seeing an opportunity to seize an easy victim, does so. There seems no doubt on the other point, that the birds and rabbits are drawn to watch the spectacle out of curiosity, and that this is strong enough in this instance to overcome the aversion to a natural enemy.

The recognition of a sense of curiosity in animals is of relatively recent date. It is only two decades ago that it was bracketed with the search for food, the need to escape enemies and the impulse to reproduce, as the four most important 'instincts'. And pride of place was given to the sense of curiosity. It may be arguable how far this was justified, but at least it can be said that the sense of curiosity is a powerful factor in behaviour even if it is not the most powerful.

This appreciation sprang from observing monkeys being kept in captivity for study purposes. It was seen that a monkey kept in a room with small windows would leave its food in order to go to a window to see what was going on outside. Deliberate experimentation to test this further led to monkeys being placed in boxes, each box having a small door, or window. A monkey so confined would open the window merely to see what was happening outside. There was no stimulus to evoke the action, no sound to attract its attention, and the only conclusion to come to was that a monkey will open a window and look out, merely to satisfy itself.

At about this time a story was going the rounds of a scientist who took a chimpanzee into a room and then went out and shut the door behind him. The scientist then bent down and put his eye to the keyhole, to see what the chimpanzee was doing. All he saw was the eye of the chimpanzee, curious to see what the scientist was doing. Whether the story was apocryphal or not is hard to say. At least it illustrates the point about animal curiosity.

◁ An animal like the Red fox that is hunted intensively, and has been hunted for centuries, is bound to show a greater alertness as compared with an animal that is not hunted by man. Part of this alertness will be directed towards becoming extremely familiar with its habitat, if only as part of the process of keeping watch for a possible attack.

There is a closely related form of animal behaviour bearing so near a resemblance to curiosity that the two are difficult to separate. Indeed, curiosity may be nothing more than an advanced or specialized form of it. It is known as exploratory behaviour. This is so commonplace that we take it for granted. A dog taken for a walk in a wood will have a wonderful time running around and about with its nose to the ground investigating everything that lies in its path, mainly through its sense of smell. There may then come the moment when it picks up the odour of something below the surface and it starts to dig. If a dog were capable of framing words the digging would be preceded by the thought: 'I wonder what is down there'. It then proceeds to try to find out. Exploratory behaviour passes at that moment to curiosity.

Exploratory Behaviour. Exploratory behaviour is more or less universal throughout the animal kingdom. Even a sedentary animal such as a Sea anemone, fixed to one spot, will wave its tentacles in the water, exploring in search of food and incidentally keeping a constant tally of the environment. If conditions in the environment change for the worse, by a sudden change in temperature, the introduction of toxic chemicals, or vibrations in the water denoting possible danger, the anemone abruptly withdraws itself.

Any free-living animal habitually carries this further by virtue of its being able to move around, to search for food, to explore a territory for a multiplicity of purposes such as noting escape routes or demarcating a boundary to that territory, seeking a mate or merely for want of something better to do. Exploratory behaviour is a form of curiosity and it must, as a phenomenon, merge insensibly into curiosity pure and simple, yet there are fairly distinct boundaries between the two phenomena.

For one thing, a sense of curiosity connotes a certain level of mental capacity and is therefore found mainly in animals with well organized brains. This means it is confined mainly if not entirely, to vertebrates, especially mammals, although it is not, as we have seen in the case of the Muscovy ducks, entirely absent from birds.

Exploratory behaviour is more certainly linked with memory than is mere curiosity, and if any hard-and-fast line is to be drawn between the two, it may be that a sense of curiosity operates on a transitory basis. It may even be that it is the pacemaker for exploratory behaviour. A specific example

may perhaps underline this. The example chosen concerns a tame genet I took over when it was a year old. It was installed in a small room that had been prepared for its occupation. There were boughs arranged around the room, from floor to ceiling, the genet being largely arboreal.

The animal's first action on being liberated in the room was to jump up at the closed window in an attempt to regain its liberty. It dropped back to the floor and after a brief pause started slowly to investigate the room, with senses alert. It made a complete circuit of the room infinitely slowly and with extreme patience, testing the floor or the bough beneath it with each foot before placing the foot firmly down. Throughout this tour its eyes were scanning, its nostrils twitching and its ears moving forwards, sideways and backwards, and also twitching. Each protuberance on the branches, each mark on the wall of the room, was examined minutely with nose and eyes. Having completed the inspection of its new quarters it started again to follow the same route, hardly less slowly than before, and following exactly the same route, even to closely inspecting each protuberance and each mark. At the end of it, and with hardly a pause, it followed the same route but this time at a brisk pace. Thereafter, the genet could race at speed round the room without mishap even in darkness, as if during the initial three inspections it had registered indelibly the details of the room.

Later, an enclosure, a large wire-netting cage the size of the room itself and similarly furnished with boughs, was erected outside. When the window was opened the genet could pass at will from room to cage. The first time it was allowed to do so it followed the same procedure of inspection of the cage as it had when first released into the room. Similarly once this had been done, it was able to race round the cage at night in darkness, sure-footedly.

The important clue to the idea that the animal registered the details of its environment lay in one episode. As the genet was inspecting a particular bough in the cage, it missed its footing and swung under the bough and had to hoist itself back onto it. For a long time, whenever it reached that point it swung under the bough and hoisted itself back onto it.

This series of observations suggests that nocturnal animals such as the genet use the original explorations of the habitat or home range to establish a guide for future forays. Diurnal animals probably

This Rhesus monkey could be testing a twig with its teeth out of curiosity or it may simply be chewing a familiar food item. Curiosity is a difficult behaviour to define or to recognize. For example, in 1907 a tame monkey escaped from captivity on an island in the Thames. It spent some time ravaging the neighbouring gardens, inspecting all the flowers and finally swam the Thames to visit the racecourse, apparently attracted by the excited crowds. It is not easy to determine what urge took the monkey across the river, but when a tame monkey puzzles for two hours over the problem of a lock and key or learns by its own investigations the secret of a screw, this seems to come nearer a pure sense of curiosity. It is also an example that seems to bridge the gap between idle curiosity and purposive investigation. In 1954 Robert A. Butler showed that Rhesus monkeys have a fundamental sense of curiosity. They will persist in examining any objects left lying around, scratching, bending or biting them, 'monkeying around' with anything they can get their hands on. Dr. Butler claimed from his experiments that curiosity, which was normally dismissed by the behaviourists as an instinct, has made the learning ability of primates possible, and is as important as hunger and sex in their success.

do much the same thing, although there is less need for memorizing since they have the greater advantage of being able to see more clearly.

In the natural sequence of events the inspection, or exploration, is built up gradually. For example, the hedgehog is strongly nocturnal. When a young hedgehog leaves the nest in which it was born it learns about its environment in two ways. It is taken out on forays by the mother and can be seen with its litter mates walking close to the mother or even following her in single file. It also carries out exploratory ventures on its own. The first time out, it walks around the nest never venturing far from it, but investigating the terrain all the time with its nose. The next night it does the same, but ventures on a longer radius. With each succeeding night it enlarges the circle until one night it goes off and does not return.

This is the basis of the exploratory behaviour in the young mammal of whatever species and it needs little imagination to visualize that it is fundamental to everyday living. As time passes the basis is enlarged until the animal is fully familiar with its territory or its home range. Further exploratory behaviour is concerned with filling in the details of everyday living, with the search for food, the location of shelters, the establishing of escape routes and the like.

Curiosity. A sense of curiosity differs from the exploratory instinct in being not linked with memory of familiar things but with the unfamiliar. Sometimes it will have the quality which in human beings we call idle curiosity, the examination of trivia merely for the sake of not allowing a thing to pass unexamined even if it is of no importance. At other times there is more purpose in the examination, as when the dog in a wood digs and possibly unearths a nest of baby rabbits. There is also the almost hypnotic arousal of interest as when birds and rabbits gather round a fox charming; the curiosity then being so strong as to inhibit the powerful instinct for self-preservation.

Whoever first promulgated the idea that a sense of curiosity was not only one of the four basic requirements in the life of an animal but also the greatest, was only correct if he included exploratory behaviour as part of it. The more compulsive sense of curiosity can be shown to dominate the other three. The birds gathering around a fox playing are temporarily subordinating the instinct for self-preservation. The monkey that leaves its food to look out of the window of its room is subordinating the instinct for acquiring food, if only briefly and temporarily. Even more striking is the way in which curiosity can overcome the third instinct – the need to reproduce.

The act of coition is normally regarded as one of the most compulsive in the life of an animal, symbolizing the urgent need to provide for the survival of the species. A male rat about to mate has been seen to abandon temporarily his amours to walk over and investigate an unusual happening a short distance away. This may not be the rule among rats or any other species, but if it happens to one rat it can happen to others.

There is a similar observation but in a different context for the Bat-eared fox. This is an aberrant species of fox. It has very large ears and it has

The Giant West African rat *Cricetomys tambianus* investigating cocoa pods. It belongs to the family Muridae, to which the Common rat belongs and, like all members of that family, spends much of its active time investigating objects both useful and useless to it. It is because of this acute sense of curiosity that rats appear to have a high ability for problem solving. Under experimental conditions in the laboratory, rats 'fiddle' with everything they meet, and, if presented with a number of levers, only one of which can be pressed to obtain food, a rat will soon learn which lever to hit for a reward.

A group of young hedgehogs after they have started to feed independently. Any one of them on its own would shuffle around so making itself familiar with its environment and at the same time indulging in exploratory behaviour.

numerous small weak teeth, its diet being mainly insects. The Bat-eared fox ranges over the arid, open plains of eastern and southern Africa. It is nocturnal but often basks by day at the entrance to its den. On once occasion a family party of Bat-eared foxes was seen to trot up to, and run beside, a hyaena carrying one of her cubs, and seemingly showing an interest in the proceedings. When the hyaena disappeared into her new den the foxes peeped down as if anxious to know what was taking place. The hyaena is a killer in addition to being a scavenger and one would have thought foxes of any kind would give even a female carrying her cub a wide berth. Moreover, there could be no profit gained by the foxes by trotting beside the hyaena or peering into her den. Rather, there was an added risk to them from a powerful carnivore jealous of the safety of her cub. Of further interest is the fact that Bat-eared foxes are said to have the habit of approaching human habitations and showing the greatest interest in what its occupants are doing.

Curiosity or Aggression? This prompts two speculations – idle speculations, perhaps! The bad reputation of the wolf is being redeemed by modern researches in the field which suggest that this animal is less aggressive than stories from the past would lead us to believe. Could it be that the wolf has the same sense of curiosity as the Bat-eared fox? It is easy to imagine that the wolf, known as a killer of large animals, and whose eerie howls could be heard in the night, was attacked when it approached houses. This would lead to retaliation by the wolf and counter-retaliation by man snowballing to open war between the two species.

The second speculation concerns the domestic dog. The weight of current opinion is that the dog is descended from the wolf. Another of several other theories is that it developed from an ancestral species other than the wolf that is now extinct in the wild. In either case it could be that a sense of curiosity drew the ancestral dog to frequent the vicinity of human dwellings and that domestication

followed from this. This would mean, if the wolf were the ancestor, that wolves were tolerated in one part of the world and not in another. It would not be the only instance of a wild animal, even a carnivore, being judged dangerous in one area and living at peace with man in another. It is true of crocodiles and leopards, to quote only two.

This hypothesis may be wholly at fault, yet the fact remains that among the Canidae a sense of curiosity is more apparent than is normally credited. The Red fox has been persecuted for centuries. Its normal reaction to man is to flee forthwith, to bolt into cover and from thence to peer out, watching intently everything the human intruder does. This could be interpreted as a reasonable means of keeping watch on the enemy except for what happens when the populations of foxes build up to unusually high proportions. This happens occasionally in Britain. Then, it is noticeable that the foxes become particularly bold. In one year when foxes were more numerous than usual, in southern England, people reported foxes approaching close to houses and squatting for long periods on end watching somebody working in the garden, apparently showing interest in all that was happening. In the same year several people reported having had a fox walk at the same pace the other side of the hedge, seemingly unafraid and apparently curious.

The puma or cougar, one of the large cats of America, also has a bad reputation and has been persecuted. Bounties have been placed on it. It has been trapped, hunted with dogs and shot. Nobody denies that it will kill domestic stock, especially ponies, but its reputation for evil goes so far beyond this that the puma is regarded as a menace to human safety. Yet there are only one or two authentic records of a person having been attacked by a puma unprovoked. Those naturalists who know the puma well, all testify to a peculiar feature of this American cat. They refer to the puma's overwhelming curiosity. It will follow a man, at a short distance, and every time the man looks round the animal melts into the undergrowth. This gives the impression that the puma is stalking the man, which is doubtless unnerving to many people. But the naturalists are categorical that it is no more than an indication of the puma's strong sense of curiosity.

All writers are unanimous that another group of the Carnivora are inquisitive or have a sense of curiosity. This is the seals. Like the Red fox, seals have been persecuted for centuries and, like the fox, they tend to make for safety before indulging their sense of curiosity in areas of the world where they have experienced the heavy hand of man. In a typical example, a man in a boat approaches a beach where the seals are hauled out. They watch him as he draws near and then make for the sea. In a short while one seal after another surfaces with just its head above water. The occupant of the boat finds himself surrounded by a circle of seals all gazing intently at him, and continuing to do so for a longer time than he has patience to wait. It is well known that seals can be attracted to a boat by music from a mouth organ or an accordion. This is more likely due to a sense of curiosity than to an appreciation of music.

A truer picture can be obtained from the account given by Captain Cook, in 1785, when he wrote about walruses in the Bering Strait. They had not been over-persecuted then and it was a bare 40 years since the walruses there had been shot at. He recorded vast numbers of walruses following his boats and coming up to them, but the moment a musket was pointed at them they submerged. Other explorers of arctic waters have given similar accounts almost to the present day, of walruses swimming out to a ship as soon as it dropped anchor, swarming around the ship and diving under it to come up on the other side. One of the dangers to a small boat or a kayak is that a walrus is likely to hook its tusks over the side and capsize it. There seems to be no suggestion that this is deliberate aggression, only that it seems to be the walrus' method of coming alongside to take a closer look.

Too often inquisitiveness is mistaken for aggression, as in the otter, where it is not persecuted. From some of the Canadian lakes, where men seldom penetrate, come reports of an otter following a canoe from one shore to the other, swimming alongside and repeatedly rearing its ferocious-looking head over the gunnel to see what the strange visitors are doing.

A sense of curiosity, or inquisitiveness, is far from being the monopoly of carnivores. The pronghorn antelope of North America has many peculiarities. One of these is its large eyes which are said to give it a particularly wide-angle range of vision, and sight equal to that of the human eye aided by x8 binoculars. It also has the reputation of being inordinately curious and a trick used to

bring it within range of the hunter's rifle was to push a stick into the ground and tie a white handkerchief on it. The pronghorn would then approach to about 40 yd (37 m) from the stick as if drawn by its own inquisitiveness.

Another peculiar feature of the pronghorn is its ability to erect the hairs of the two white patches on its rump when alarmed or excited. Then, as the animal bounds away, the white patches flash like a heliograph, warning other pronghorns that there is danger around. It could be argued that a white handkerchief, fluttering in the breeze, could be sufficiently like the flashing white patches of another to deceive a pronghorn into trying to catch up with it. It is hard to accept such a proposition, since an animal with x8 vision should not need to come so close to satisfy itself of the nature of the handkerchief. Secondly, under persecution the pronghorn has become shy and wary. It does not approach a decoy as closely as formerly, which suggests inquisitiveness rather than the instinct to follow the white rump of a congener. Thirdly, an

In sharp contrast to the apparent inquisitiveness of the Common or Harbour seal, the Elephant seal seems to be without any particular sense of curiosity. In this picture a female is giving the 'open mouth' threat display at an intruder. She will soon tire of this. Meanwhile, her fellow elephant seals, lying farther away from the camera, are just not interested enough even to raise their heads.

The Orang utan has been hunted by man for food since the early Stone Age. Since the advent of the shotgun this ape has been even more intensively persecuted. This may partly account for its shy and retiring nature, as a result of which an orang will take cover when it sees a human observer. Some individuals are more tolerant of being looked at by man and it has been said that this may readily be transformed into intense curiosity if a human has become familiar and shown himself to be friendly. Orangs also become very curious about their environment when they are in captivity.

alternative trick, which seems to have worked well as a decoy, was to persuade a boy to lie on his back and wave his legs in the air.

Similar hunters' tricks have been reported for other ungulates, and the sequence of events follows closely a basic pattern for all species. It begins with the animal drawing near to inspect man. Soon, persecution makes them more wary, after which the inquisitiveness continues but at a greater distance. There is, however, at least one exception, or presumed exception. This is the Antarctic wolf, a fox-like member of the Canidae living in the Falkland Islands. When Europeans first landed there they were greeted by these foxes coming to meet them. The foxes looked like wolves, inspired fear and were shot. Even so, others continued to approach settlers so close that they could be clubbed. The Antarctic wolf became extinct before it could learn to sub-ordinate its sense of curiosity to its need to protect itself from enemies by keeping at a safe distance. The sense of curiosity is undoubtedly linked with a higher organization of the brain and is most strongly developed in the Primates, the monkeys, apes and man. In the last of these we title it idle curiosity, a spirit of enquiry or scientific research, according to circumstances, although all three are merely successive stages of the same mental process. In all species, men included, the exploratory behaviour is instinctive and much used, with the sense of curiosity as a specialized expression of it. The exploratory behaviour is most marked, in man, in the first years of life and is epitomized by the actions of the developing infant. Indeed, mothers are prompted to declare of their infants, that they 'have fingers into everything', when in fact they are merely exploring their environment. The sense of curiosity develops at a later stage and it is most marked in the brighter individuals. In fact it is probable that it underlies adaptive or evolutionary success, rats and men providing two outstanding examples. It may, indeed, be that it is in this manner that the sense of curiosity can justifiably be regarded as the most important of the four fundamental instincts.

With several studies made of chimpanzees in the field, our knowledge of these apes has been much increased in recent years. Until then, study of chimpanzees in captivity had revealed a lively curiosity and a readiness to examine closely unfamiliar objects.

20

Food and Feeding

'Greedy pig', term of contempt for anyone with a ravenous appetite, who will eat almost anything he can lay his hands on, is an insult to the pig. In its natural state, or even under good farm conditions, this much maligned beast will no more over-eat than will any other animal. A pig's only fault, if fault it be, is that it will eat almost anything edible, plant or animal. It will relish roots and stalks equally with leaves or fruits, and it will eat any animal it can catch or any carrion it finds. Its propensity for digging up underground plants is exploited in locating truffles, the delectable fungi that grow a foot down in the ground. It is a tribute to a pig's acute sense of smell that it can find them.

Animals that eat a wide variety of foods habitually and without discrimination are known as omni-vores, which merely means eating everything. This is in contrast to herbivores, that eat only plant food, and carnivores, the flesh-eaters. Pigs are not the only omnivores, nor the only animals with hearty appetites. One virtue of being an omnivore is that conditions have got to be very bad before the animal feels the pinch of starvation.

Because its flesh is palatable man has domesticated the pig to supply him with food, and he has tended to wipe out the wild pig for both food and, later, sport. But left to themselves pigs flourish because they are never at a loss for something to eat. To be an omnivore is to start with an initial advantage in terms of survival.

At the other extreme are those animals with a restricted diet. These are relatively few. Those that

Although principally a vegetarian, the wild boar eats leaves, berries, roots and tubers, and especially acorns, it readily turns to animal food, from earthworms, insects and reptiles to birds and their eggs. It will also readily take carrion.

have a restricted diet can exist successfully and in large numbers provided there is an abundance of their particular food, but their survival is threatened as soon as there are adverse changes in their environment. The prime example among mammals is the koala, the so-called koala bear, of Australia. The koala, in no way related to bears but to kangaroos, eats mainly the leaves of eucalyptus (gum) trees. Moreover, it eats only certain leaves of particular kinds of gum trees. That is why we see no koalas in zoos outside Australia.

Before the white man settled in Australia there were plenty of gum trees and plenty of koalas. It is true that several decades ago koalas were killed in millions for their fur, but the greater menace to them has been the cutting down of gum trees in the course of human settlement of eastern Australia, the main home of the koala.

Today, when Australians would like to bring back the koala from the verge of extinction, the feeding of this animal poses a problem. It is not sufficient to set aside an area of land as a national park for its preservation. There must be in such a reserve a sufficient number of well-grown gum trees of the right species. Such areas are not easy to find.

Other specialist feeders include the Three-toed sloth and the Red spruce mouse of northern California. The first of these feeds exclusively on flowers and leaves of the *Cecropia* trees. The Red spruce mouse eats nothing but the needles of the Douglas fir.

Perhaps a word of caution should be registered here. Our knowledge of foods and feeding habits is only moderately complete for a few mammals. Even in these revision of our ideas is necessary from time to time as more information is gleaned. The Giant panda, for example, was formerly thought to eat only bamboo shoots. Then it was found it also ate in the wild fishes, small birds and rodents, as well as other plants. At the London Zoo the Giant panda, *Chi-chi*, was eventually fed on a mash of rice, fruit, sugar, milk and eggs, to which was added roast chicken and vitamins. The panda also enjoyed blackcurrant jam and spaghetti.

In a completely natural state the fate of any species of animal with this kind of narrowly restricted diet hangs by a thread. Any natural catastrophe, such as disease, which wipes out or seriously reduces its food supply, spells extinction, because the animal cannot suddenly switch to another source of food.

Tastes and preferences. Comprehensive surveys of diet, as has been said, have been made for relatively few mammals. When such surveys are made, they are based mainly on examination of the stomach contents of dead individuals of a given species supplemented at times by an analysis of the droppings. The results are then expressed in tables showing the range of foods eaten by that species. Such surveys are valuable but they are sometimes misleading because no allowance is made for individual preferences. Thus, anyone claiming to give a complete list of foods eaten by foxes in Britain would not fail to include poultry. Yet those informed on the subject agree that poultry-stealing by foxes is localised. A vixen who is a poultry thief will instil in her cubs, by feeding them with chickens,

The diet of the koala was formerly considered to be extremely restricted. Some reports said that populations in different areas fed on one species of eucalyptus in each area and confined their attentions to certain leaves only of that one species. It is now known that any koala will feed on a number of species of eucalyptus, even occasionally eating the leaves of other kinds of trees. Even so, for a herbivorous mammal its diet is outstandingly narrow.

When the Giant panda was first brought to the London Zoo it was firmly believed by all zoologists that it ate nothing but bamboo shoots. At that time a public appeal was made by the zoo authorities to anyone growing large clumps of bamboo in their gardens or parks to help maintain the supply of this exotic plant to feed the panda. It was not until some years later that Chinese zoologists observing the animal in the wild noted that it not only ate other plants but even small animals as well. In recent years it has been found that the Giant panda could be fed successively on a variety of foods including roast chicken!

a taste for poultry. On the other hand, there are many instances of foxes living beside poultry farms feeding exclusively on rodents, supplemented by wild fruits.

The leopard is another animal against which there seems to be a mistaken prejudice. Leopards are supposed to be dangerous to domestic animals and humans. This, apparently, is by no means always true. It seems more likely that they have pronounced individual feeding habits. One lived near a farm, in country abounding in small antelopes, yet it was seen to eat only bush-pigs. Moreover, its droppings never showed any other than pig bristles. Another fed largely, if not entirely, on fish (*Tilapia*) caught with the paw while the leopard crouched over the edge of a bank at the lakeside.

One leopard was the subject of extended observations by a farmer, who was at first inclined to shoot it. He desisted when he realized the leopard was preying on the baboons that regularly ravaged his crops. It never once molested his livestock and his dogs were unharmed, although leopards have the reputation for preferring dog to any other prey.

Cannibalism is not unknown among leopards and may represent another individual preference. Feeding habits also tend to vary with the age of the individual. Young leopards and very old individuals account for large numbers of small prey such as Cane rats, hares and ground birds.

It may be that the bad reputation of leopards as a species needs the kind of corrective already applied to the reputations of wolves and hyaenas. Wolves it now seems are not naturally aggressive to man in spite of the blood-curdling stories that have been written about them. Even more drastic has been the overthrow of the belief that hyaenas are cowards battening on lions for the remains of their kill. That they are scavengers is still true. It is recorded that in the 18th century the Cape butchers in South Africa gave free access to slaughterhouses and the like to the hyaenas that regularly came at night to clean up offal, bones and skin.

In recent years Hans Kruuk of the Serengeti Research Institute followed hyaenas on moonlit nights in a Land-Rover. He found they made their own kills of zebra or wildebeest and by dawn had devoured everything, leaving not even a splinter of bone.

Kruuk also found that as often as not it was the lion that waited for the leavings of hyaena kills, instead of hyaenas waiting for the remains of lions' kills.

The hyaena's ability to crunch even long bones of large animals and to digest them is one of the seven wonders of animal feeding habits. The ability to crack large bones depends on strong teeth and powerful jaws. The reason usually given for hyaenas and dogs being able to gulp chunks of flesh and pieces of bone, and suffer no digestive disorder, is that their digestive juices are particularly strong. Even in the human stomach the presence of hydrochloric acid, one of the most corrosive of acids, has long posed the puzzle how the digestive juices can break up meat without damaging the wall of the stomach. The complete answer has yet to be found but there may be a clue in the fact that the lining of the stomach sheds and replaces large numbers of cells. In the human stomach this replacement is at the rate of half a million cells a minute.

A specialist feeder, like the koala, suffers not only from its inability to turn to alternative foods but also from its failure to have developed the habit of hoarding. Gum leaves may not lend themselves to hoarding. Certainly many mammals, herbivores, omnivores and carnivores, make use of this.

Hoarding and storing. Food hoarding is most common among rodents; the typical example in everybody's mind is the squirrel. Some species of squirrels, the North American chickarees for example, store great heaps of pine cones under damp earth. Grey squirrels, also North American, bury nuts or acorns singly at all points of the compass. Most mice make hoards. The Longtailed field mouse of Europe is especially zealous in this, laying in large quantities of berries, acorns or nuts. In fact, their stores are often ascribed to squirrels by the people who uncover them.

One of the more striking hoarders is the pika, or Mouse-hare, of North America and Central Asia. It anticipated human haymaking, cutting grass and other herbage in the summer, laying it out to dry, then storing it under rock overhangs for winter use. The Bandicoot rat or Indian mole rat stores large quantities of grain in its burrows and it is said that local villagers will go to the burrows to collect these in times of famine.

The Wood rat, also known as the Pack rat or Trade rat, of North America, hoards food and other objects. Its peculiarity is that it leaves something in exchange. The classic example is of the box in a mining camp from which a supply of metal nuts were removed and replaced by pebbles. A more amusing instance of hoarding gone mad concerns the Brown rat, in Britain, which filled its nest with 'three towels, two serviettes, five dustcloths, two pairs of linen knickerbockers, six linen handkerchiefs, one silk handkerchief, . . . 12 lb (5·4 kg) of sugar, a pudding, a stalk of celery, a beet, carrots, turnips and potatoes'. The action of the rat could

be interpreted as an exaggerated form of gathering nesting material as well as food. The reason for the Pack rat's behaviour is less easy to understand.

So strong is the hoarding instinct in rodents that a House mouse living in a larder will accumulate a miscellaneous pile of foodstuffs beside its nest although surrounded by more food than it could consume in its lifetime.

Among carnivores, Arctic and Red foxes, the fennec, wolf, coyote, Black-backed jackal, bears and weasels are known to store surplus food. The fox will dig a hole with a forepaw to bury a mouse, pushing earth back over the cached food with its nose. Other carnivores use a less pronounced form of caching food. A puma, having fed, will cover the remains of its kill with brushwood or other vegetable rubbish and will return to it daily until all is consumed. Tigers also cover the remains of a kill. The Arctic fox will cache surplus lemmings in summer, pressing the carcases into crevices between rocks. When leopards tree their kill it is not caching but a means of keeping the carcase out of reach of lions. Leopards living in areas where there are no lions do not tree their kill.

There have been reports of weasels killing more than they could consume and hoarding the carcases of mice in a remarkable way. First a line of carcases is laid out neatly in a row and covered with earth. Then a second row is laid on top and this in turn is covered with earth. The precision of this operation, as described by observers in the earlier years of this century has aroused scepticism. There was an occasion, however, when a household cat found a nest of well-grown baby rats. It killed one, carried it 12 ft (3·7 m) along a garden path and laid it on the path. Then, in leisurely fashion, returned to kill another, carry it and lay it neatly beside the second. This was repeated until eight dead baby rats lay in a neat row side-by-side. Then the cat lost interest. In the light of this, the habit attributed to the weasel does not seem so far-fetched.

Mole rats are reported to bite the growing points out of corms and tubers, before storing them. This may be because they like the growing points, but the effect is to prevent the corms and tubers from sprouting and losing their value as stored food. Kangaroo rats of the deserts of the southwestern

The hyaena is a typical carnivore but with jaws and teeth ▷ much more powerful than usual. For a long time it has been believed that the hyaena lived entirely on the leavings of other animals' kills. Recent study has shown that it will do its own hunting, but in addition, it is prepared to eat any carrion.

United States store seeds in pits an inch deep near their burrows. They cover these with sand. Only when they have dried out, so cannot germinate, are they taken into the burrow to be stored. More remarkable than this is the procedure adopted by the European common mole in storing earthworms.

During the early years of this century a controversy raged, through letters to the Press, about the origin of stores of earthworms dug out of the ground by gardeners and others. Some people argued that the worms had been brought together by moles. Others disputed this. The matter was settled by a Danish zoologist, M. Degerbøl, in 1927, in the only rational way possible. He live-trapped moles and made available a superfluity of worms. The moles satisfied their appetites, then they bit the heads off the worms and cached them. The advantage of this is that a worm lacking its first few segments cannot burrow so it remains a prisoner but still alive and healthy. Should worms elsewhere remain in normal supply the worms regenerate their head ends and escape, so living to be eaten again another day.

Short-tailed shrews, relatives of moles, are said to store snails and beetles, biting them to prevent their escaping while remaining alive to give the shrews a supply of fresh meat. Shrews are unusual among mammals in having a venomous bite. It has been suggested that venom may paralyse the bitten victims.

Hunters' over-kill. Storing food is a matter of prudence rather than gluttony. Most animals feed only to satisfy the appetite. Overkilling springs from impulses other than greed, although what these impulses may be is not always obvious. Several of the smaller carnivores, especially those belonging to the weasel family, have been branded for 'killing for the sake of killing', a misleading phrase. A European badger may kill a dozen ducks or domestic hens and eat only one. Foxes will do the same. It is usually supposed that a badger doing this is a 'rogue', by which is meant a sort of animal delinquent that departs from normal behaviour. For foxes, one explanation is that a multiple killer is a young one. There may be a better explanation derived from observed behaviour of domestic dogs 'worrying' sheep.

When two or three dogs free in the countryside get in among a flock of sheep they may savage not one but many. These are not necessarily young dogs. The greater likelihood is that as soon as one bites a sheep it engenders a group excitement. Even one dog on its own, unused to sheep, has been seen to start killing the moment the sheep begin to stampede, whereas on other occasions, when the sheep have stood their ground, there have been no casualties. It is as if the hunting instinct is roused not by hunger alone, but needs to be stimulated by the chase. The cheetah illustrates this. It picks out its quarry from among a herd of antelopes, then dashes at top speed – variously estimated at 50–70 mph (80–113 kph) – to seize its victim. If it fails to overtake its prey in this overwhelming rush, the cheetah stops, squats and loses interest. It has also been observed that if its quarry stands its ground, refusing to be panicked into running, the cheetah on drawing near checks its speed stops and turns away.

The question of surplus killings has recently been surveyed by Hans Kruuk, both from his own observations and those of others. Foxes have been known to go through a colony of nesting Black-headed gulls killing or maiming 4–10 per cent of the total number of adults and eating only parts of a few of them. In a Sandwich tern colony 12–15 per cent were killed and only 17 per cent of these were consumed. Rabbits are on occasion treated simil-

A pile of bones in a hyaena's lair in Kenya suggests that the animal may have a tendency to hoard food, much as a dog will bury a bone. In this wide assortment are some human remains which were probably exhumed by the hyaena from a neighbouring cemetery.

Wolves are adapted to killing large prey and their normal method of hunting is the pursuit of fairly large land animals. Like the domestic dog, whose ancestor it is believed to be, it will reject no form of flesh, living or dead, including fish.

arly by foxes. Surplus killing of caribou calves and of White-tailed deer have been recorded for wolves. Spotted hyaenas have made surplus kills of Thomson's gazelle and of wildebeest calves, and similar kills are known for leopards with goats, lions with wildebeest, Polar bears with narwhals and Hunting dogs with warthogs.

Nocturnal predators make surplus kills when the prey is thickly congregated and, more importantly, when the night is dark, both moonless and with an overcast sky. That is a time when the prey makes no attempt at flight, due doubtless to a feeling of security in the blackness. In other instances, where there has been a surplus kill by day, the prey

animals were penned in, naturally or artificially. For example, when they were hemmed in by ice floes.

In all these instances the passage of the predator has been marked by a trail of dead or disabled prey, with only small indications of their flesh having been eaten. The percentages of prey attacked are always similar, ranging from 4 to 15 per cent. The biological significance is not clear, since if this were at all general the predators would succeed only in destroying their own food supply, which is what would assuredly happen. No species can survive for long a five per cent loss over and above the normal wastage. This is why human predation is so disastrous, especially since the introduction of

27

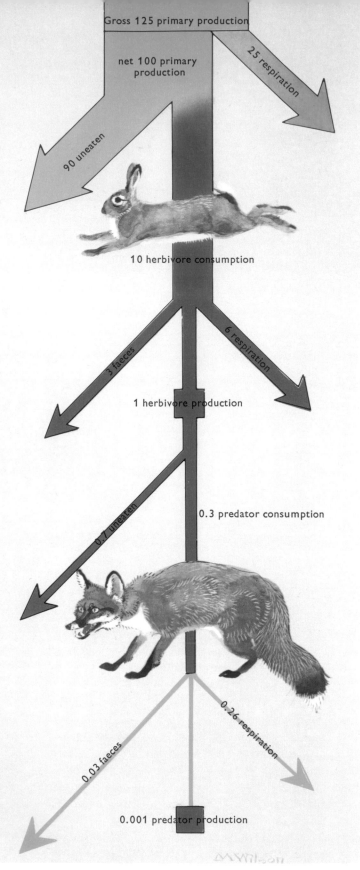

Gross 125 primary production

net 100 primary production

25 respiration

90 uneaten

10 herbivore consumption

3 faeces

6 respiration

1 herbivore production

0.7 uneaten

0.3 predator consumption

0.03 faeces

0.26 respiration

0.001 predator production

M Wilson

Diagram illustrating the main energy flow through an animal community. Of the gross primary production of vegetation 25 per cent is lost in respiration, 90 per cent is uneaten. Of the 10 per cent eaten by a herbivore only 1 per cent is available to its carnivorous predator whose primary production is in turn reduced by 90 per cent due to wastage.

firearms, and more particularly when the killing has been wanton, as happened with the quagga, for example.

The reason for surplus killing is hard to find. It may perhaps be found in the familiar circumstances, that people who have to work hard for food do not waste it. Those to whom food comes easily squander it. Normally, for a hunting animal there are four stages: searching, hunting, catching, killing. It is when the first two are eliminated, by the passive behaviour of the prey that surplus killing occurs.

Role of the predator. It has long been accepted that the predator benefits its prey species by weeding out the sickly, weakly and aged, and more especially by taking the young, thus helping to keep the population within reasonable bounds. A species that exceeds its normal numbers tends to destroy its own habitat, and with this its own food supply. This can result in starvation and disease, and a catastrophic fall in numbers – a population crash. The cheetah's behaviour illustrates well the first way in which a predator keeps a stock healthy.

As has been observed on many occasions for predators such as lion, leopard and wild dog, the cheetah appears to select as victim a member of the herd that shows signs of weakness or sickness. Should its judgment be at fault, and it selects a healthy individual, its quarry outstrips the cheetah in speed and the pursuit is not pressed home. Remarkably, also, if the quarry shows determination by not fleeing it is likely to be left alone.

Some years ago, an experienced naturalist, remarked that 'there appears to be something cowardly in the predator'. This comment was based on the number of times a predator, with the advantages of weapons (teeth and claws) in its favour, has turned away from a relatively defenceless quarry that has turned on it to fight for its life. Whether this is a matter of cowardice, or merely of prudence, it shows that the predator does not always have things its own way.

There are other ways in which the predator can be frustrated. Hugo and Jane van Lawick-Goodall watched Cape hunting dogs in 91 chases, of which only 39 were successful. If this represents a typical pattern it is not surprising that pregnant females, newborn babies, the weak, sickly and the aged should be the main targets of attack. Without them the predator would even more often go hungry or be forced to turn to smaller or more menial victims. And there are many recorded instances in which this has happened.

The leopard is a typical carnivore and this tends to make us assume that any leopard will naturally prey on warm-blooded animals to the exclusion of everything else. In fact the leopard seems very much an individualist in the matter of diet, but unfortunately its habits have not been studied in the wild.

How much food? As has been said, animals do not normally overeat, although there is one, above all others, that is reputed to do so. It is the largest member of the weasel family, sometimes known as the wolverine, but also and more commonly as the glutton. It is known by this second name in French, *glouton*, in German, *Vielfrass*, in Dutch, *Veedvrass*, and the scientific name is *Gulo gulo*, the Latin for greediness being *gulo*.

The alternative name of the wolverine is based more on the wide diet of the beast, which will eat almost anything animal, including carrion. It also has the reputation of robbing fur-hunters' traps, and there is a legend about this large member of the weasel family, which is nearly 33 in (0·8 m) long in the body, with only a six-inch tail, and weighs 33 lb (15 kg). The legend is that it will gorge itself until it can eat no more, then squeeze through a narrow space between rocks to empty stomach and intestine in order to eat more. A disgusting thought and wholly unjustified. Nevertheless, it shows what people formerly thought about it

Much of the beliefs in former times about animal nutrition had something of this exaggerated quality. It expressed itself in the apparent eagerness with which authors have declared that this or the other animal eats more than its own body weight of food daily. Small mammals especially were prone to be credited with this propensity.

The smaller the animal the greater its surface area relative to bulk and the more it loses body heat by radiation. As a consequence, small mammals are liable to suffer from 'cold starvation'. That is, they must eat to replace not only the energy used up in searching for food but also the loss of heat from the unusually large surface of the body. Peter Crowcroft has put forward an interesting theory why shrews have so often been said to have these enormous appetites. Naturalists, he suggested, caught shrews and took them home alive to keep them in captivity for study purposes. By the time the shrew was installed in its cage and given food, it ate ravenously. A naturalist seeing this took the first opportunity to weigh such a shrew and also the food it ate during a

29

To the casual observer, this British red squirrel appears to pick up a nut in its front paws and hold it to its mouth to eat it. More careful observation shows that the squirrel is unable to pick up food and that it first takes the nut in its teeth, rises on its haunches, cups its forepaws and drops the nut into them. This is clearly an ingrained action and contrasts with the skilful manipulation of the front paws in opening a tin box in order to steal biscuits, bread and other foods.

given period of time. By a simple essay in arithmetic he estimated that the unfortunate prisoner ate one and a half times its body weight of food in 24 hours.

More sober analyses in recent years have shown that shrews, and other small mammals that have been supposed to consume their own body weight or more a day, may sometimes do so but more commonly they eat only 75 per cent of their body weight. Moreover, it has been found that even these conservative figures must be qualified. The bodies of earthworms and insects, especially insect larvae, which form the staple diet of many small mammals, comprise a high water content. If only dry weights of food were considered the bulk per day would be significantly less than even the 75 per cent already mentioned.

It is the herbivores, the grass-eaters and vegetarians, we think of as taking in an excessive bulk of food. No animal excites people's curiosity about the amount it eats than the elephant, with its enormous paunch. It is a strict herbivore, and although so familiar an animal nobody has yet succeeded in measuring with any degree of precision the amount consumed by an elephant in the wild. All we have is reasonable guesses that it is between 336–672 lb (152–305 kg) a day. This estimate is coupled with the comment that probably a large bull elephant may eat twice as much as the largest of these amounts. Since a large bull may weigh up to six tons, this sets the elephant's daily consumption at little more than one-sixth of its body weight. This despite the fact that it will be feeding sixteen hours out of every twenty-four.

Even the assessments for zoo elephants can only be made in general terms. Richard Carrington summarized that a full-grown individual will be fed 100 lb (45·4 kg) of hay a day supplemented by bran, oats and root vegetables. And zoo elephants take little exercise!

Digestion. No animal achieves complete digestion of the food it eats. Some of the smaller animals, such as Dung beetles, may come very near to doing so. The majority void faeces that still contain nourishment for a wide variety of other organisms. In this, the elephant affords an outstanding example. In spite of its huge crushing molars it squanders about 50 per cent of its food in the form of imperfectly digested boluses.

Francis G. Benedict showed this by an ingenious method. He cut up a rubber inner tube of a car tyre, in distinctive shapes, and placed these pieces in loaves of bread fed to zoo elephants. If the elephant with its huge molars were efficient in chewing this must have had its effect on the rubber. Yet, the pieces of rubber were recovered from its faeces showing no tooth-marks. Benedict further showed that only 44 per cent of the hay was digested. This compares with 50–70 per cent for cow, sheep and horse.

The figures for sheep and cow are bound to be higher than for elephants since they are ruminants. That is, instead of a single-chambered stomach there are four compartments to the alimentary tract between the gullet and the intestine. The first is a capacious rumen. A ruminant crops grass or leaves, twigs and bark. When grass is the main ingredient the animal is said to graze. The animal that eats chiefly leaves is said to browse.

A ruminant crops grass or leaves and swallows them incompletely chewed, but moistened with a copious supply of saliva. In the rumen the food is attacked by anaerobic bacteria which break down the cellulose in it. Having filled its rumen the ruminant then retires to cover, seeks shade or merely rests, according to the species. Then follows the process known as cudding, or chewing the cud.

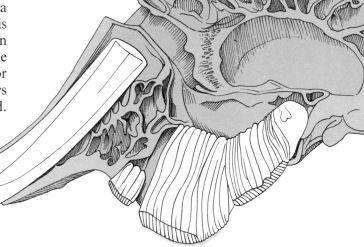

Elephant's teeth consist of huge molars for grinding plant food. In the course of its lifetime an elephant has seven molars on each side of both upper and lower jaw.

The food is brought back to the mouth in small boluses, one at a time to be thoroughly masticated by the flat-crowned molars acting as grinders as the lower jaw moves in a sideways chewing motion. When fully masticated, the food is swallowed again. It passes through the rumen again, but this time it goes direct, assisted by a valve, into the second compartment, the reticulum, thence into the psalterium, on to the abomasum, and into the intestine. The series of digestive processes as the food passes from one to the other of these organs ensures a high degree of digestion.

Dolphins and porpoises, supposed to be descended from land herbivores but now fish-eaters, also have a four-chambered stomach but there is no rumination. The stomach compartments are markedly dissimilar, also, from those of a ruminant. The first compartment is a crop, little more than a wide dilatation of the gullet. This is followed by three other, much smaller compartments, with glandular walls, before the intestine is reached. Since fish has a higher food value than grass these animals are able to do without the large compartments seen in herbivores.

In many herbivores auxiliary digestion is provided by a caecum, a long blind tube opening from the intestine where the small intestine merges into the large intestine. At its blind end is a small fleshy tube, also blind, the vermiform appendix. In the human the vermiform appendix is present without a caecum and is that part which is removed surgically to cure appendicitis. In the rabbit the caecum is long and wide, and plays at least as important a part in digestion as the large and small intestines.

Just over 30 years ago a second auxiliary digestive process was discovered in the European rabbit. It was first called 'chewing the cud', but although it is in principle like the process used by ruminants there is a marked difference in execution. As a consequence it was later called refection. The process had been noted in the European hare as early as 1881.

Refection is the habit of passing food twice through the intestine. Domestic rabbits swallow their night droppings without chewing them. Wild rabbits swallow certain of their droppings twice daily. The droppings refected are different from the faecal pellets proper. They are soft, lighter in colour and are mainly from the caecum, where they have been partially digested by bacteria. It is believed they contain large amounts of vitamin B_{12} and vitamin deficiency can occur if the animal is prevented from refecting.

Since it was first described for the rabbit and hare, refection has also been recorded for the European common shrew and several rodents. Time will doubtless show that the habit is extremely widespread among rodents and insectivores. Further study may also reveal that in all animals practising refection, as has been so far proven only in rabbits, the process is important for survival of the young. Young rabbits while still suckling eat the soft faeces of the mother, so introducing into their digestive systems the intestinal bacteria which break down cellulose. If prevented from doing so the baby rabbits die of convulsions.

Feeding the young. Refection is one of several ways in which the offspring draw essential substances, in addition to food as such, from the mother. One of the first acts of a newborn mammal, and one that takes place very early, is to explore. The exploratory behaviour leads it to its mother's teats, to feed. It can hardly be that hunger pangs are driving it since it will have been supplied with food from the mother through the placenta practically until the moment of birth. Its first drink of milk does more than nourish it. The first milk the baby

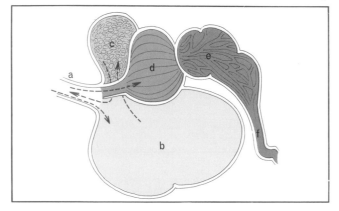

The multiple stomach of a ruminant: (a) oesophagus, (b) rumen, (c) reticulum, (d) psalterium, (e) abomasum, (f) duodenum. Dotted lines and arrows show how the food, when first swallowed, passes into the rumen and to some extent into the reticulum, is then regurgitated, fully masticated and swallowed again, passing through the reticulum into the psalterium and abomasum.

draws from the mother's mammary glands is known as colostrum. It differs from the milk it receives after the first day or so of its life. Colostrum sets the physiological components of the digestive system to work in a co-ordinated manner. Among other things it acts in the manner of a purgative, cleaning out the intestine. In hoofed mammals it contains substances that help the young animal to build up resistance to disease. In other mammals this resistance is built up either before birth, or after birth but independently of the colostrum. The importance of the first feeds of mother's milk for the young hoofed animal is seen in the fact that if artificially reared from birth it has little chance of surviving to adulthood.

All baby mammals are nourished on mother's milk – that is part of the definition of a mammal. The length of time they do so, and the richness of the milk they obtain, vary from one species to another. The time varies from three weeks, in some of the shrews, to five months in the larger hoofed animals. As a rule, the young animal starts to chew solid food before it is weaned. Baby guinea pigs will

chew grass when no more than two to three days old. This precocious chewing in young herbivores may be more a matter of learning to use the teeth than of obtaining nourishment because at first the chewed food is not swallowed. It may represent a natural apprenticeship against the time when the mother's milk gives out and the youngster must feed itself.

Young carnivores have food brought to them by their parents. Thus they are initiated into the appearance of potential prey as well as being encouraged to chew, so reinforcing any natural instinct to try feeding themselves. Young herbivores find their food all around them. It does not have to be brought to them. Moreover, the parent does nothing, so far as is known, to help it select the correct food from the various plants available. Perhaps the infant herbivore must rely on its own trial-and-error testing. Perhaps it starts to chew in imitation of the parent and before the real need for it arises. These are a few of the many questions to which at present there are no firm answers.

It may even be that the taste-buds play an important role in helping the young herbivore to choose its food. The taste-buds are groups of sensory cells on the surface of the tongue and in the lining of the mouth. In the human tongue they are grouped for taste: sweet on the tip of the tongue with salt either side of it, bitter at the back of the tongue with areas of sour on either side. A wine-taster sips because the buds for sweetness are on the tongue tip. A beer taster drinks to take the liquid to the back of the tongue where the buds for bitter taste are located.

Less is known about the distribution of the taste-buds on animal tongues, but counts have been made of their numbers in several species. It is significant that herbivorous mammals have more taste-buds, as a rule, than those that eat insects or flesh. A bat, an insect-eater, has 800 taste-buds, the human tongue has 9,000, omnivores such as pigs and goats have 15,000, a European rabbit has 17,000 and an ox 35,000.

There is one striking anomaly. The hare, so like a rabbit in appearance and habits, and closely related to it, has only 9,000 against a rabbit's 17,000.

The European Harvest mouse feeds mainly on seeds, and is ▷ able to climb slender stems of herbaceous plants in search of these delicacies.

Living Without Drinking

Food and drink are indissolubly linked in the human mind, but whereas the first often contains the second, water – frequently adulterated in the human diet – is the natural drink of wild animals and is solely concerned with the balancing of water loss by water gain. Food on the other hand provides the essential materials for growth and energy.

In vertebrates water is lost from the skin and lungs, in urine and faeces and, in marine fishes by osmosis through the gills and body surface. It is gained by drinking, eating food which contains water, by oxidation of food otherwise called metabolic water and by osmotic uptake through the skin in freshwater fishes. A water balance in the body is needed to keep the tissues normal, to provide an adequate volume of blood and, in mammals, for temperature regulation. We drink to quench a thirst, which is the body's signal that the water balance is upset and more water has been lost than is good for us.

Man, like his near relatives the monkeys and apes, is basically a vegetarian and it has been suggested that he first took to eating the flesh of other mammals when he found himself in a desert environment. Whether the theory about early man is true or not, it can serve as a striking illustration of the fact that the higher the water content of the food the less the need to drink. This has an important bearing on the lives of animals, particularly of mammals. There are, for example, large desert-living mammals, such as the gemsbok, that can replenish their daily water loss by eating desert succulents. These are plants able to store in their tissues the water absorbed by their roots, instead of, as in normal plants, giving water off continuously to the atmosphere.

Except in desert or semi-desert, water is usually always available and therefore taken for granted. Consequently, little is recorded of the drinking habits of animals apart from those living in deserts. One that has been intensively investigated is the camel, largely because of the legend that it stores water in its stomach. This was an absurd idea if only for two reasons. The first is that water stored in the stomach, or even one compartment of the stomach, must seriously interfere with if not wholly jeopardize normal feeding. The second is that water, like alcohol, can pass straight through the wall of the stomach.

The legend, now shown to be a myth, of the camel's water storage depended upon two things. One was the finding by an early anatomist that the wall of one of the three compartments into which a camel's stomach is divided is lined with row upon row of small pockets. It was assumed the water was stored in these. The second was the accounts, given by several explorers in the 19th century, of finding themselves without water in the blazing heat, slaughtering their camels and drinking the contents of the stomach.

It is now known that these pockets are glandular, secreting digestive juices. Anyone who has had to go without water for several days, even in a temperate climate, knows the desperate craving for liquid that can ensue and how, at last, when driven to drink it, even foul water can taste akin to nectar. Moreover, one's judgement becomes warped, so the word of these travellers should never have been accepted at its face value.

The liquid in the pockets lining the camel's stomach is an evil-smelling soup of liquefied, masticated food. It could be drunk by a man crazed with thirst who in this condition would probably be past judging whether he was drinking liquefied herbage or merely foul water.

The full story of the camel's physiological adaptations to desert conditions is remarkable and more will be said about them later. In the meantime we turn to those of some other desert dwellers that are if anything even more remarkable, if only because the animals are able to go indefinitely without water and do so without eating moist food.

The Kangaroo rat. There may have been reports of other animals capable of doing this, if so they were probably treated with scepticism, but the one that really caught the interest of scientists was the Kangaroo rat of the deserts of the southwestern

A herd of vicuña drink at a stream. Vicuñas are related to ▷ camels and live at heights of up to 16,600 ft (5,000 m) in the Central Andes. At such heights, there is an abundance of water in the 'bofedalts'—streams fed by melting snow.

United States. Some of these were kept in captivity and it was noticed they could live indefinitely on a diet of dry seeds.

The Kangaroo rat was thoroughly investigated about 20 years ago by Dr Bodil Schmidt-Nielsen, in the United States. She found that the Kangaroo rat is not independent of pre-formed water, that is, free water as distinct from water derived by oxidation of the food inside the body, which is known as metabolic water. Nevertheless, a Kangaroo rat can go indefinitely without drinking. In their desert habitat rain falls only at very long intervals and dew is not normally formed, so to live without drinking is compulsory. It is accomplished by two means, by making the maximum use of such water as is available and by using this water with the utmost economy.

The food of the Kangaroo rat is seeds and these take up a certain amount of water. The rodent stores seeds in its burrows in quantity. The inside of the burrow, being a few inches below the surface, remains constant at a temperature below 86°F (30°C) and the absolute humidity is 3–4 times as high as it is outside the burrow. The Kangaroo rat is nocturnal and by spending the daytime in its burrow there is little or no evaporation through the skin. Moreover, there is virtually no loss of moisture from the exhaled breath partly because, in the burrow, the air is humid and partly because, at all times, the moisture in the exhaled breath is condensed in the nasal passages, where the temperature is 50°F (10°C) lower than that of the rest of the body.

A Kangaroo rat has no sweat glands. The kidneys have a concentrating capacity nearly five times greater than that of the human kidney, so the Kangaroo rat's urine is highly concentrated. Its faeces are dry and hard. In addition there is refection. That is, some of the faecal pellets are swallowed allowing the rodent to recover the small amount of contained moisture as well as vitamins formed by bacterial action in the intestine but not absorbed by the intestinal lining.

So much for the conservation of water. But there must be some water loss, however small, and therefore some replenishment of water. This comes from two sources. The first is from pre-formed water taken in by the seeds stored in the humid burrows. The second is from metabolic water formed, after the seeds have been eaten, by oxidation.

There is positively no storage of water in the

The Common eland in East Africa, in the dry savannah. During the dry season elands will dig up bulbs with their hoofs and eat melons and other succulent fruits, in a manner recalling the behaviour of other large antelopes such as the gemsbok.

A herd of Asiatic cattle, known as the zebu or Humped cattle, also much used in tropical Africa. They have advantages over cattle breeds of the temperate regions, which are also imported to the tropics. They will graze and rest under a fierce sun when Western cattle need to seek the shade. They eat less at a time than European breeds but feed more often. They are better suited to life in areas of poor grazing land where the grass is dry and sparse. They need water and will take full advantage of an opportunity to drink, but they can go for longer periods than European cattle without it.

sense of special storage spaces or organs, or pockets in the tissue where free water is lodged.

In captivity, Dr Schmidt-Nielsen found the Kangaroo rats given access to water would ultimately learn to drink and they would take fresh or salty water readily, and thrive equally well on either.

This form of water balance, and the ability to survive without drinking, is found in some other desert rodents, and can be suspected for others. The Pocket mice of the United States, which with the Kangaroo rats form the family Heteromyidae, also have this ability. So have the gerbils, jirds and Desert rats (family Cricetidae) and the jerboas (family Dipodidae) of Africa and Asia and the Kangaroo mice (family Muridae) of Australia. All agree in build and habits, as well as in their physiological adaptations.

These rodents furnish a good example of convergent evolution or possibly, to use a more recent term, of adaptive convergence. All are nocturnal, desert-living, with long hind legs for bounding over sand. And all have the ability to go without drinking.

Some of the larger mammals have this same ability. Some years ago a naturalist reported having followed an Arabian oryx for a week or so without seeing it drink. This received little notice on the part of zoologists, probably because they doubted the truth of it. The finding has now been confirmed. The addax, an antelope four feet at the shoulder, living in the Sahara, seems as able as a camel to tolerate desert conditions. It is a shy beast so its physiology has been little studied, but it seems also to be able to go without water indefinitely.

Even today relatively little can be found in books on the drinking habits of mammals, probably because it is assumed that animals, like ourselves, drink daily. It is a surprise to learn that the African hunting dog, and probably also the jackals can go without drinking for as much as two months.

Hyaenas also seem to be able to go for long periods although they will drink every day, if water is available, and will also wallow in water, as well as mud, both of which would help the water balance of their bodies. In none of these has there ever been any suggestion that they store water in any part of the body, although some of them can go for longer periods without drinking than have been recorded for the camel.

It has sometimes been said that an elephant stores water although nobody seems clear on where or how it does it, except for vague suggestions that it may be stored in the throat. The origin of this idea may lie in the elephant's habit of blowing out a fine damp spray at intervals of moisture that has condensed in its trunk. It may increase this by putting its trunk into its mouth and withdrawing a certain amount of mucus.

It is not surprising that this idea should have been generated if only on the grounds of simple arithmetic. Thus, an elephant drinking first sucks up about $1\frac{1}{2}$ gallons with its trunk then puts the tip of the trunk well into its mouth and squirts the water down its throat. A herd of elephants will visit a river or a waterhole in the early morning and their drinking sessions last for up to half an hour, seldom less than 10 minutes. The enormous intake of water is, however, more an indication of the wasteful use of water by its body, especially since there is normally, for each day, a second drinking session in the evening. In hot weather there may be a third visit at midday.

The problem of living in a hot desert is not so much the intensity of heat by day as the scarcity of water and food, the two being closely linked. The lack of water compels an animal to eat less, the Bedouin have found this also. Perhaps the converse is true. Certainly among humans heavy eaters are almost invariably heavy drinkers, and ascetics drink only in moderation. If so, it could indicate a connection between the enormous meals of an elephant and its daily intake of water. Nor is there any question of an elephant storing water. With such a large surface area there is the opportunity to lose considerable quantities of water, whether through sweat glands or by insensible evaporation, that is, moisture lost through the skin and in the exhaled breath, the latter being at its maximum when an animal pants.

Another form of insensible evaporation is seen in the use of saliva. The Kangaroo rat is compelled to avoid the heat of the desert day because, by an anomaly, it cannot stand heat. In laboratory tests it was found that kangaroo rats die within the hour when exposed to temperatures of 100°F (38°C), which is only slightly higher than the human body temperature and none survived an exposure of $1\frac{1}{2}$ hours to 104°F (40°C), temperatures which are far lower than those they would experience by day in their natural habitat. They can, however, bring in another mechanism for temporary relief, by dribbling saliva.

This has its counterpart in what is called the kangaroo's water bag, when a kangaroo licks its wrists copiously. The evaporation causes a cooling of the skin at that point, which means body heat is being lost. It is an interesting question whether the licking seen as part of the daily behaviour of

Camels drinking at a well near Darawar in Pakistan. So long as water is available a camel will drink like any other animal. It is, however, able to go without drinking for long periods and then drink copiously when water is available.

The camel is the epitomy in most people's eyes of an animal that can live without drinking. There are others which can do this more successfully owing to a remarkable series of adaptations aimed at the conservation of their body fluids. However, the camel can go for long periods in a completely arid environment, abstaining from drinking and allowing its body to become dehydrated, yet recovering fully from an emaciated condition once it reaches a water supply.

some mammals, and included under the heading of grooming, may not be partly due to an instinctive method for keeping cool. It may also account for the hedgehog's self-anointing. In this the animal licks an object, accumulating a foamy saliva in its mouth, then contorts its body to place the saliva with its tongue on various parts of the spiny covering to its body. This may be repeated until the hedgehog is more or less coated with its own saliva. Various explanations have been offered for this strange behaviour, including that it may be a perverted form of grooming.

Were Hedgehogs once Desert Mammals? A cactus is a plant adapted to a desert life by having leaves reduced to spines. It is a fanciful idea to see in a hedgehog a primarily desert animal that has its body covered cactus-like with spines. There is a species of desert hedgehog in Africa, although most other species distributed over Europe, Asia and Africa live in less inhospitable habitats. Because

hedgehogs are nocturnal and secretive, remaining immobile or rolling into a ball when approached, little is known of what they do on their nightly foraging. How much or how often they drink can only be a matter of conjecture. Many people in Britain put out bread-and-milk in the evening for the hedgehogs living in their gardens. Hedgehogs in captivity will daily consume a similar ration. But is this merely opportunism, a matter of drinking when liquid is available, as in African hunting dogs, hyaenas and jackals and, more especially, wild asses and camels?

This seems to be the only interpretation to be placed on an episode that occurred in southern England several years ago. The owner of a garden found a hedgehog one day apparently sick, if not moribund. The following morning he filled a low bird bath, as a matter of daily routine. He always filled a pint jug to do this. A few minutes after the bird bath had been filled the 'dying' hedgehog

39

tottered over to it and drank it empty. Since the body of a hedgehog is less than a foot long the animal had drunk water equivalent to about half its own body volume. Having done so, the hedgehog seemed to be fully revived and ran speedily away.

A feasible explanation could be that the hedgehog was suffering from dehydration and might have succumbed but for the timely arrival of a supply of water. The normal diet of European hedgehogs is earthworms, slugs and insects, all having a high water content. Dehydration could ensue if food was scarce, as in a dry spell.

The behaviour of this hedgehog almost exactly parallels that of the wild ass and the camel, both desert animals. The ass can go without water for long spells and then replenish its tissues with a long drink, taking in up to 15 gal (68 l) of water in five minutes. A camel can do the same, but it can imbibe up to 30 gal (136 l) in 10 minutes. During that time the camel will pass from an emaciated animal, showing all its ribs, to a normal appearance except that the body is swollen on one side.

The Camel. It was Pliny, the Roman naturalist (AD 23–79) who first recorded the idea that a camel stores water in its stomach. Buffon and Cuvier, celebrated 18th century French scientists accepted it and the British anatomists Owen and Lyddeker, in the 19th century, supported it. Captain A. Leese, veterinary surgeon with camels for many years, cast doubt on it in his book published in 1926, and Schmidt-Nielsen showed it to be incorrect.

In fact, the camel does not store water any more than does any other animal. What it can do is to survive dehydration of the tissues to a remarkable extent and then hydrate them again when water is available for it to drink large quantities. A man losing 12 per cent of his body water, by sweating, evaporation and excretion is in dire distress because the water is drawn from all his tissues, including his blood. This becomes thick and turgid and the heart has difficulty in pumping it. A camel can lose 25 per cent of its body water and still show no signs of distress, because the water is drawn from all its tissues other than the blood. So there is no strain on the heart and an emaciated camel is still capable of normal physical exertion.

It is sometimes stated that the camel stores water in its hump. This contains fat and it has been argued that this can be metabolized to yield water, therefore the hump must be a water reserve. The hump of an Arabian camel may contain as much as 100 lb (45 kg) of fat, each pound of which could yield 1·1 lb (0·5 kg) of water. This would amount to 13 gal (59 l) for 100 lb (45 kg) of hump. To convert the fat to water would, however, require extra oxygen. It has been calculated that the breathing needed to make available the extra oxygen would lead to the loss of more than 13 gal (59 l) of water as vapour in the breath. A camel, so far as is known, does not possess the capability seen in the Kangaroo rat of condensing the vapour, in the exhaled breath, in its nostrils.

Many of the accounts given of the camel's ability to keep going without water can be discounted.

The probability is that hedgehogs in the wild obtain the necessary liquid from their food, which consists of earthworms, slugs and insects all of which have a high water content. Like the camel the hedgehog seems to drink copiously when the opportunity is there and is probably able to go without drinking for a significant period, provided it can restock its tissues in due course.

The Indian blackbuck lives on the plains of peninsular India. It conserves water by feeding during the morning and late afternoon and lying up during the heat of the day, a common pattern of behaviour among mammals that live in hot and arid environments.

Nevertheless, the authentic records are sufficiently remarkable. There is one of a march through Somalia lasting eight days, without the camel's drinking. In Northern Australia a train of camels made a journey of 537 mi (864 km) lasting 34 days. Most of the camels died but a few that were able to graze dew-wetted vegetation survived. They might have done better had there been bushes for them to browse, judging from recent research on the feeding of African antelopes.

During a drought in Africa, domesticated cattle suffer badly whereas the indigenous antelopes survive. Cattle eat grass almost entirely, the roots of which are shallow, lying in the top few inches of the dried out soil. Bushes are more deeply rooted and are able to draw upon stores of water deeper in the soil. Added to this the leaves of bushes are hygroscopic so are moist at night, when the humidity of

the atmosphere is relatively high. This explains why antelopes do most of their browsing at night. Even during a drought, leaves of the acacia may consist of 58 per cent water at night. It has been shown that an eland browsing at night takes exactly the amount of water needed to keep its body weight constant.

Antelopes, including the eland, have another advantage. They are able to let their body temperature rise by several degrees centigrade during the day without sweating and so conserve an appreciable amount of body water; in an eland weighing half a ton the amount conserved is five quarts. As compared with cattle, also, the urine, faeces and milk of antelopes are more concentrated.

Methods of drinking do not vary much in mammals. Most of them lap up water with the tongue but the ungulates typically suck up water, even the giraffe and the camel with long necks have no difficulty. The elephant's way of drinking has already been described but it is worth mentioning at this point that an elephant can find where there is water just below the surface and can bore for it, using either one of its tusks or the end of the trunk, in the loose substratum of a dry river bed. The rhinoceros is also able to tap sub-surface water, and the pools so formed by rhinoceroses and elephants are afterwards visited by smaller animals in times of water shortage.

The Dwarf mongoose of Africa will sometimes drink by lapping but more usually it dips a forepaw in the liquid and then licks the paw.

There are a few exceptional methods of drinking among mammals. Most if not all insectivorous bats drink on the wing by skimming just over the surface and dropping the lower jaw to scoop up a small drop of water. It is a delicate operation because if the jaw were dropped too far it would create a drag that would pull the bat into the water. The result need not be fatal because bats have been seen to fall into water by colliding with an obstacle just above the surface and then take off from water or swim to the shore by flapping the wings.

More remarkable still is the method used by the Lattice-winged bat of Trinidad. This feeds on soft fruit pulp and juice using what might be described as a natural suction and filter. Indeed, the bat could be said to be using a liquidizer. It feeds on over-ripe bananas and pawpaws. Although it has teeth these are not strong, only sufficient for tearing open the fruit. The skin between bat's lips and gums is covered with numerous fleshy papillae and with mouth half-closed the bat sucks up the fruit juice and mushy pulp, with the papillae straining off any large pieces that could not pass down its throat, which has a diameter of only a twelfth of an inch. Behind the throat is the gullet which leads into another tiny opening, and behind this is a kind of sac or bag leading from the gullet. The method of use of this unusual swallowing apparatus is not known precisely, but its effect seems to be to convert the food into meat and drink in one operation.

Too Much Water. Having explored, if in a cursory manner, some of the ways in which animals live without drinking, and comparing these with animals which do drink, we can suitably end the chapter with a brief look at those mammals that spend all or most of their lives in water. Many mammals are called aquatic even though they enter water mainly for catching their food. The otter is an example. The three groups that are truly aquatic are all marine. They are the Sea cows, the dugong and manatees, the cetaceans, which include whales, dolphins and porpoises, and the seals. The latter spend part of their lives on land. There is, in addition, the Sea otter, of the North American Pacific waters, but so little is known of its day-to-day biology that we can only speculate on its drinking habits by analogy with the other marine groups.

The Sea cows live in coastal or estuarine waters, the manatees sometimes ascending rivers, feeding on aquatic plants which are themselves largely made up of water. Moreover, living almost wholly submerged as they do, at most only occasionally rearing their heads and shoulders out of water, insensible evaporation is minimal.

When water is available giraffes will drink daily, bending their long legs to reach the water or straddling them. Giraffes can, however, survive in dry regions where there is no free water. It seems they can satisfy their needs from the juices contained in the foliage they consume.

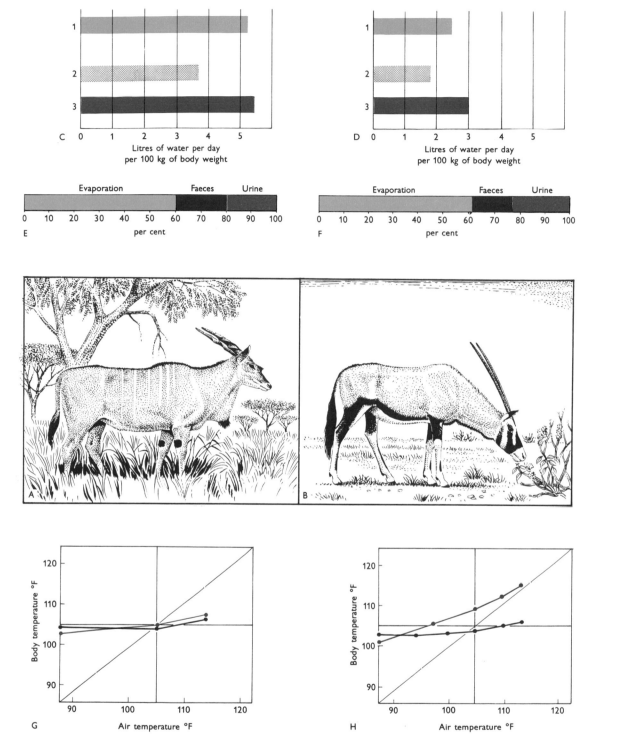

Some mammals are able to survive with less water than is required by others. Here, two African antelopes, the eland (A) and the oryx (B) are compared. The eland lives in moderately arid regions where over 5 litres of water per 100 kg of animal body weight are available each day (C1). Of this, just under 4 litres are required by the eland under normal conditions (C2) and over 5 litres under hot conditions. The amount of water available to the oryx is far less (D1), but the oryx requires correspondingly less water under normal (D2) and hot (D3) conditions. Both animals require more water under hot conditions because, in both, about 60 per cent of their body water lost is lost through evaporation from the skin (E and F). When an eland is subjected to hot conditions (G) its body temperature falls below the temperature of its surroundings when the latter is above 104°F (40°C). It does this by sweating whether it is dehydrated (red), or has abundant water (black). The oryx (H) is better suited to hot conditions because, when it is dehydrated (red), it can allow its body temperature to rise with that of its surroundings. It does not need to lose heat by sweating, and so can save water.

Seals, it appears, obtain all the water they need from their food, usually fish or cephalopods (squid and octopus). The walrus eats molluscs, the Crabeater seal eats the plankton animals known as krill, the food in both being comparable with fish and cephalopods for water content. Little sea water is swallowed with the food. There are freshwater seals, the Caspian and Lake Baikal seals. These sometimes drink water and are said to be able, as a consequence, to go for longer periods without food than seals living in the sea. According to Judith King, an authority on the subject, seals are normally not seen drinking although those in captivity have been seen to drink freshwater and salt water.

Zebu or Humped cattle originated in India, and spread to China, Africa and even South America. Their popularity in these parts of the world is explained by the fact that they can withstand hot climates better than can western cattle breeds. They are able to maintain a low body temperature in hot weather because their skin contains a greater number of sweat glands than is normal in cattle. It has also been suggested that the zebu stores fat in the hump clearly seen in this photograph. Concentration of fat in one place makes it possible for the zebu to avoid having a fat layer covering its entire body, an obvious disadvantage in hot climates.

In former times elephants were found throughout what are now the temperate and Arctic regions of the Northern Hemisphere. Their modern survivors are found in tropical countries where it would seem they are ill adapted to live, except through their behaviour. The daily life cycle of a wild elephant is one of fairly continuous movement in search of food and water, the hottest parts of the day being spent in the shade.

Remarks about the cetaceans must be even less dogmatic, but it seems highly probable that they do not drink at all. They share with the Sea cows almost total immersion at all times, so insensible evaporation will be minimal. The Whalebone whales feed on krill and must swallow some water with their food, which itself has a high water content. The Toothed whales, including dolphins and porpoises, like most seals, feed mainly on fish and cephalopods, with a similar effect to that in seals. The exception is the Killer whale which feeds on other cetaceans, seals, sea birds and almost anything else it can snap up. It can only be presumed that, as with other cetaceans, it swallows some water; and its food also has a high water content.

There may be some insensible evaporation through the breath, as there must be in Sea cows and seals, and this brings to mind the whale's blow. It has always been said that this is caused by moisture in the breath condensing on meeting the cold air as the whale surfaces and breathes out. But the 'blow' is as conspicuous in tropical as in polar seas. Clearly there must be another explanation. The spray of the blow is, in fact, made up of droplets of mucus from the lining of the respiratory tubes. Its function is not known for certain.

Whales are descended from land mammals, probably from ungulate stock, possibly from the same ancestors as elephants. It may or may not be pure coincidence that the elephant blows out a spray of mucus.

Even if the spray is made up of droplets of mucus these must contain water and therefore represent a water loss, however small relatively to the animal's size. Whales also lose water through their intestine as a consequence of maintaining the salt balance of the body. Like all other mammals the females lose water when nursing the young and it is of interest to see how the water content of whales' and seals' milk compares with that in other mammals. It is highest in the milk of camels and asses, in which the water is 87·7 and 90·3 per cent respectively and lowest in the common dolphin, whale and seal with 44·9, 54·8 and 46·4 per cent respectively. In those other mammals for which a milk analysis is available the percentages of water lie between these two ranges. Regrettably the analysis for the Kangaroo rat is not available.

Hunters and Hunted

Some animals are born to be hunters, others are liable to be hunted. The first are called predators, the second are the prey. There is one order of mammals, the Carnivora or flesh-eaters, that recommend themselves immediately as the arch-hunters, the killers, the animals that most inspired the now outmoded phrase 'Nature red in tooth and claw'. But for the fact that the larger of these sometimes kill people we would always have looked upon them dispassionately as playing an essential part in the balance of nature. Instead, man has branded them as ferocious and blood-thirsty, and everywhere has sought to exterminate them. Some of the large carnivores, such as lion, tiger and leopard, are at times a menace to human life. So man's wrath has descended on them. The smaller carnivores, such as fox, wild cat and weasel have become a menace to poultry and to game preserves. So these have been persecuted. Too often, however, both large and small carnivores have been killed for sport.

In killing off the carnivores man has let in the pests. In Africa baboons are a menace to crops where the leopard has been killed off. In India deer have become an equal nuisance in the absence of tigers and other big cats. Europe has had plagues of rats which foxes, wild cats and polecats might have kept in check had they themselves not been killed off.

The most distinctive feature of the Carnivora is their teeth, the structure of which underline their method of feeding. They have the usual types of teeth found in mammals, the incisors in the front of the jaws, followed by canines, pre-molars and molars. The last premolar of the upper jaw and the first molar of the lower jaw are modified to form long, sharp ridges that can shear flesh like a pair of powerful scissors. These two pairs of teeth are known as carnassials. The canines also are large and powerful and are used in slashing blows in killing prey and also in fighting.

The order Carnivora is divided into eight families as follows: Canidae or dog family (dog, wolf, fox, jackal); Felidae or cat family (cat, lion, tiger, leopard, jaguar, lynx, cheetah); Mustelidae or weasel family (weasel, stoat or ermine, badger, otter, skunk); Procyonidae or panda family (panda, raccoon, coati, kinkajou); Protelidae (aardwolf); Hyaenidae (hyaena); Ursidae (bear); Viverridae (mongoose, civet, genet).

This is a mixed bag of animals ranging from the Kodiak brown bear, which may weigh up to 1656 lb (751 kg) to the weasel weighing 1·25 oz (35·4 gm). In it are the aardwolf that eats nothing but soft-bodied insects, and has very weak teeth, hyaenas able to crack the long bones of large herbivores, and the pandas, coati and kinkajou that are almost vegetarians. Yet all are united by having similar teeth.

Among those carnivores that kill live prey the hunting methods vary, the most marked difference being between the typical dogs and the typical cats. All Carnivora have well-developed claws. Those of the dogs are blunt and serve to grip the ground in running and the typical dogs pursue their prey by scent or sight and run it down. Having caught up with it they weaken it with slashing blows, in which the canines play the main role, before giving the coup de grace. Typical cats have sharp retractile claws, which can be withdrawn into protective sheeths, and these are not brought into use until the moment the prey is to be seized. The method of hunting is to stalk the prey and depend on a final pounce to make the kill.

Between the typical dogs and typical cats is the cheetah, which runs its prey down, and then uses the final pounce. It has semi-retractile claws. The claws in the remaining families are non-retractile for the most part, semi-retractile in some of the Procyonidae and some of the Viverridae.

Are Carnivores Cruel? The question is not infrequently asked how far killing of animals by other animals involves cruelty. Only the victim can assess the extent of pain or suffering involved. There have been men who have been mauled by a lion, for example, have been rescued at the last moment and have lived to recount their experiences. They have been unanimous in declaring that they felt neither pain nor fear until the episode was finished and they had time to feel their wounds and recall the horror of the situation. Animal prey would stand

A lioness with her kill in East Africa. The forelegs of lions ▷ are enormously powerful, capable of breaking the neck of a zebra with one blow of a paw.

little chance of rescue, and the inevitable death blots out everything.

A film taken of lions killing a wildebeest showed the latter lying on its side with one lioness laying a paw over its shoulder. The legs of the wildebeest were moving rapidly as if the animal were trying to run. To the human eye it looked pitiful. Reason suggests that the leg movements were due to a post-mortem reflex. A lion kills by breaking the neck with a heavy blow of the paw, by gripping the nostrils and causing suffocation or by biting at the neck and severing the jugular vessel. Small prey is flattened with a blow of the heavy, powerful paw. All methods are calculated to bring speedy unconsciousness, if not instant death.

The puma, a less powerful cat than the lion, in its final violent spring has been known to knock its victim 20 ft (6 m) along the ground, before fastening its teeth in its throat.

It often happens that a domestic cat plays with a mouse near its own food bowl. The mouse being released by the cat is as likely as not to turn to the cat's food and start eating. This could be interpreted perhaps as a displacement activity like that of birds, when conflicting impulses to flee or to fight are channelled into mock feeding. In any event, it suggests an insensitivity to the menace of the moment.

A domestic cat playing with a live mouse seems to human eyes the acme of callous cruelty. Sometimes the mouse will rear up on its haunches with the front paws together as if in prayer. The human onlooker feels a wave of pity for a mouse looking as if it is begging for mercy, unless he knows that this is the aggressive posture of a mouse, and that this mouse is showing fight.

Cheetahs sometimes eat small prey alive, as do leopards and lions. It is horrible to the human observer, but everything depends on how long the victim remains sensitive to pain. The concussion

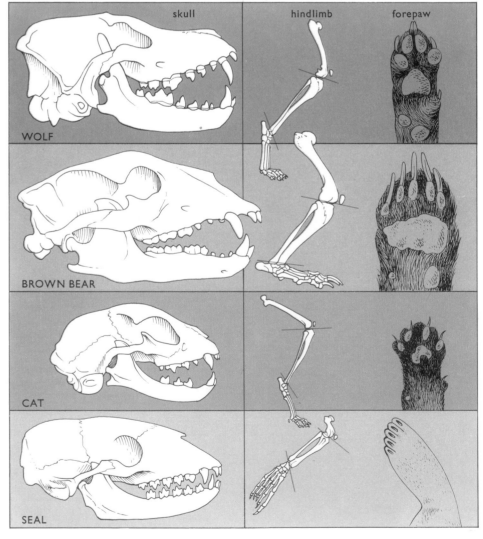

skull   hindlimb   forepaw

WOLF

BROWN BEAR

CAT

SEAL

The diagram shows the skull, the bones of the hind limb, together with the forepaw of four typical carnivorous mammals. In the wolf and the cat the cheek teeth include the carnassials, a molar in each half of upper and lower jaw which have sharp cusps that engage each other like the blades of scissors and function for slicing flesh. By contrast the molars of the bear are mainly flat crowned indicating an omnivorous diet, which may include plants, small animals and larger flesh bearing animals. The cheek teeth of seals bear pointed cusps suitable for holding slippery prey.

Cheetahs with a zebra kill. These flesh eaters are unusual members of the cat family because, instead of catching their prey by stalking with a final pounce, they run it down in one short sharp sprint. They are reputed to travel at anything up to 70 mi (113 km) per hour, a figure which probably is exaggerated. What is certain, is that if a cheetah fails to make a kill within a distance of a quarter of a mile (400 m), it loses interest and squats on its haunches waiting for another prey.

alone probably robs the prey of all sensitivity. Dogs probably cause a slower death, but this situation can best be judged from what is known among smaller animals that have fallen prey to carnivores.

One of the best descriptions of the emotions of a prey animal in the grip of a predator is that by A. Hoogerwerf. He tells of seeing a Grey macaque monkey held in the mouth of a Komodo dragon, the large Monitor lizard. The monkey, still alive but bleeding from a wound on the leg, was held in the lizard's jaws by its head but in such a way that it could look round. The monkey gave no indication of fear or panic. It could, according to Hoogerwerf, have easily scratched out the eyes of the lizard. Moreover, as the lizard walked away the monkey, instead of seeking to delay this with its feet, or offering any resistance, moved its legs as though to walk along with the lizard. It seemed the monkey was in a shock condition that inhibited any attempt to escape or defend itself throughout the twenty minutes taken to swallow it, although even when three-quarters of its body had disappeared inside the mouth signs of breathing could still be detected.

Whatever may be the final truth about the killing, the result of predation is not only to curb the numbers of the prey but to benefit the species as a whole. In controlling its numbers, predators prevent a population of animals from reaching high numbers, so causing a spoliation of the habitat leading to widespread disease and starvation. Moreover, the predation itself normally results in the killing of the weak, the sickly, and the aged. So the stock is kept healthy.

Methods of Defence. It seems firmly established now that the role of the predator is beneficial to the prey species. Another truth, slow to emerge but now becoming increasingly certain, is that the super-predators, such as the big cats as well as the other carnivores, do not have things all their own way. As they have evolved more specialized teeth and claws for the kill, so their prey have evolved greater speed to escape, as well as other means of defence or protection.

Mongooses for example, are credited with a high immunity to snake bite, a coarse fur which when raised forms a shield against the snake's fangs and a speed of movement that outmatches the snake's agility. However, even though it may be assumed that the contest between mongoose and snake is always weighted in favour of the mongoose, it emerges that the mongoose as often as not is killed by the snake.

Zebra are a favourite prey of the lion, apparently having no other defence than speed. Yet not uncommonly lions are kicked in the face by the stallion, who takes on responsibility for defence of the herd. The resulting damaged teeth forces the lion to subsist painfully on small prey and starvation can result. What proportion of lions is represented by the words 'not uncommonly' cannot be accurately assessed. It is based on the many instances recorded in the literature linked with the assumption that the majority of such casualties go unrecorded.

Another highly interesting aspect of the hunting of wild predators is also supported by observations recorded by many experienced naturalists, at least for the larger Carnivora. If a wolf, for example, pursues a moose and that animal stands its ground and looks at the oncoming killer, that killer will slacken its pace and finally come to a halt. Provided the moose continues to stand its ground and look at it the wolf will eventually turn away and seek a meal elsewhere. It is when the intended prey turns and runs away that it is in danger. Similar observations are recorded for cheetahs, lions and leopards. Whether this also happens with the smaller carnivores is hard to say.

There is another aspect of predation that seems so far to have been overlooked. To understand this it is helpful first to recall the history of the Pribilov fur seals. These seals breed on the Pribilov islands in the Pacific off Alaska. They numbered $2\frac{1}{2}$ million until sealers reduced their numbers to 200,000. They have been restored to their former numbers by making use of a principle that came to light through a study of their breeding habits.

At breeding time each adult bull holds a territory on the beach and around him the females gather, forming a harem. The young bulls, four to six years old, form bachelor groups around the periphery of the breeding beaches. They fight among themselves, causing many casualties, and they challenge the harem bulls, who race across their territories to drive them away, crushing with their heavy bodies any pups lying in their path. The American biologists charged with the task of protecting the seals, hit upon an ingenious method. First, all sealing was suspended. Then, when the numbers of the seals had recovered, they recommended the killing of a proportion of the bachelors and marketing their skins. By killing only bachelors, the heavy casualties from their fighting would be avoided and it would be possible not only for the total number of seals to increase but for a sufficient number of skins to be available for the fur trade. Today, by careful control, the Pribilov seal is back to its original numbers and 60,000 skins are collected each year from killing males that are surplus to breeding requirements.

As will be shown in a later chapter, it is the rule in many antelopes for the young males to be driven to the periphery of the breeding grounds. When monkeys move around the young males are on the periphery of the troop. One observer after another has commented on the way predators tend to seize these young males. There is therefore, a close parallel between this natural culling and man's artificial culling of the fur seals.

Similar culling of excess males appears to occur in natural situations. It is the rule that in species

Fur trade records from the Hudson Bay Company show that the population of lynx (red line in hundreds) is to a large extent controlled by the number of Snowshoe hares (green line in thousands) available as food. The periodic decline in the numbers of hares is reflected, usually a year later, in a decline in the number of lynx. The numbers of hares are linked with changes in the vegetation.

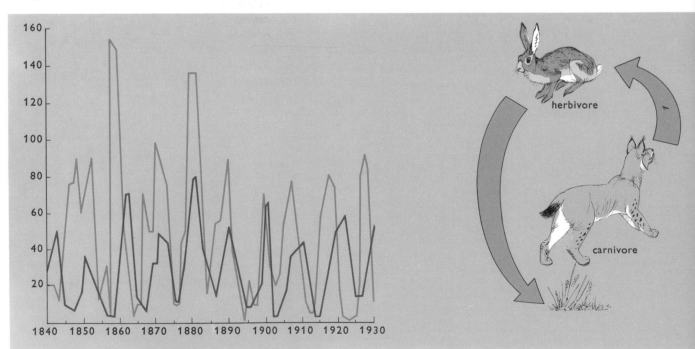

Zebras live in herds, often in association with one or another species of antelope. They are able to live in these mixed herds because the component species take different foods. So although all are herbivorous, even grass eaters, the zebra takes coarser grass than the others, so avoiding competition for food.

with a social organization where there are more than two individuals in a unit, such as troops and herds, polygamy is the rule. Since the ratios of male and female births are even this means a high proportion of the males are redundant and therefore expendible. By being banished to the periphery these fulfil a double purpose. They are in the best position to assist in the defence of the rest, including the females and young, and they also serve to feed the predators, which further reduces the inroads that would be made into the rest.

The Advantages of Community Life. There are other advantages to living together in groups. A solitary individual has to rely on one set of senses to warn it of danger. The greater the number living together the more pairs of eyes, ears and nostrils there are to detect danger, which offsets the fact that there are more mouths for the available food. As to detecting danger, it has been commonly noticed that if, for example, a zebra lies down to take a nap during the day, there will be another quite obviously standing by to give warning of the approach to danger. This seems to be so ingrained in herd animals that even in mixed herds, for example of zebra and wildebeest, if a zebra lies down to sleep a wildebeest may act as sentry for it.

51

Aggregations of individuals also produce a dazzle effect. A predator must pick out its intended victim and then await the best moment for attack. If it is one of several and they start to move around it becomes more difficult for the predator to keep its eye on its quarry, and the more there are and the more erratic their movements the greater its task becomes. This is strikingly illustrated by a description of young springbok, given by Dr R. C. Bigalke.

Young springbok form 'nursery groups' within the herd. When such a group is disturbed by a human intruder the young springbok explode in all directions, hopping about on stiff legs displaying the white patches of their rumps, which tend to dazzle the eye almost like mirrors. If a predator finds it anything like as hard to concentrate on one particular animal as does the human observer, this could have high survival value.

Adult springbok also have this 'mirror' or 'fan' of long white hairs extending from the middle of the back to the tail. They also leap about on stiff legs when disturbed and before taking flight. These characteristics may have been largely responsible for the enormous numbers in which they existed before man took a hand in reducing their numbers.

Living Close to One's Enemies. There is one form of protection that is denied to herd animals. It can be exemplified by the Red fox which lived to a ripe old age and finally died of senility. It had its earth on the boundary of a kennels in south-west England. Yet although it lived within easy reach of the foxhounds, and although the hounds and the huntsmen passed near it every time they went out to hunt foxes it was never chased. The fact is that when men go out hunting they do not start looking for their quarry immediately. So anything living on their doorstep is safe.

It is the same with animal hunters. When they leave their den or lair they do not start hunting until they have passed through their territory and into the home range. That presumably is why people have seen a fox playing with a rabbit or a young hare just outside the earth and then going off to hunt. It is why songbirds often nest below the nest of a hawk and are not molested. Anteaters may burrow into a termite nest yet leave the insect occupants unharmed. It seems also that carnivores generally follow the same rule. The more interesting fact is that wild herbivores do not crop the grass or the bushes in the equivalent of their territory but go into the home range to feed. This leaves an area of thick bush or long grass into which they can retire to rest and chew the cud, hidden from their enemies.

Generally, one has a picture of herd animals forever on the alert for dangers which beset them on every hand. Such a situation, were it correct, would lead to intolerable stress, and this would be in strong contradiction to the scene normally presented, for example by the herds of game on the African savannah.

In fact, the percentage of total losses in an animal population due to predation is surprisingly low. The general pattern of mortality is as follows: in the breeding season the population rises steeply then begins to fall steadily until by the beginning of the next breeding season the numbers are back to normal. The new generation suffers high losses from infancy onwards until only a few survive to maturity, these survivors being no more than are necessary to replace the losses among adults. For birds, careful censuses have shown that 60–75 per cent of young birds die during the first three to six months of life. The percentage will naturally vary with the species and its reproductive rate, but these figures serve at least to exemplify the normal state of affairs.

In an area where the hunted and the hunters are both present and in equilibrium, the hunters will take especially the weak, the sick and the aged, as we have seen, as well as culling the subadult males on the periphery in the case of the herd animals.

There is one saving grace in the life of a herd animal: that it seems to know immediately when a beast of prey is hunting and when it is not. A zebra or a wildebeest will move away from undergrowth or bushes if it detects the scent of lion, because it would suspect an ambush. But a lion can walk through a herd of zebra or antelopes, causing them to do no more than raise their heads or stop grazing, provided it is well-fed. The clue seems to be in the posture of the lion. If well-fed its head will be up; if looking for a kill its head would be down and it would be walking in a half-crouch. That would warn the herbivores to be ready to take evasive action.

Camouflage. How far camouflage is used in the to-and-fro battle between herbivores and carnivores will long remain a matter for speculation. Most mammals wear neutral or drab colours. They lack brilliant colours such as are found in birds and

The moose *Alces alces* is known as the elk in northern ▷ Europe and Asia. It feeds on leaves, bark and on aquatic plants. In the past, the greatest enemy of the moose was man, but it is now a protected animal.

reptiles, in which eyesight counts more than the sense of smell. In monkeys coloured patches are more common, as in the mandrill, and they are also eyesight-animals. Some gazelles, like Thomson's gazelle, has a dark band along the flank, which could act as a disruptive pattern, breaking up the animal's outline.

The most obvious examples of a coat pattern that could serve for concealment are in the leopard, tiger and zebra. Camouflage, no matter how concealing it may be, is effective only so long as its wearer is immobile. The spotted coat of a leopard would be especially effective when its wearer is lying along a branch with the dappled sunlight through the foliage falling on it. It is a favourite trick of leopards to lie up in trees and drop down onto victims passing by underneath. More commonly they stalk their prey, moving stealthily through undergrowth, where they would also be hard to see.

It is usually taken for granted that a tiger's stripes make their wearer inconspicuous in long grass because it simulates the vertical stems and their shadows. Most people who describe, from first-hand experience, the movements of a tiger on the prowl speak of nothing being visible other than the waving of the tops of the grass until the animal steps out into the clearing where the observer is standing.

The most controversial subject is the value of the zebra's stripes. People who are used to seeing zebra are agreed that in strong light they stand out starkly against any background and are much more conspicuous than the dun-coloured wildebeest that may be standing near them. In dim light, as at dawn or dusk, or in moonlight or in slight mist, the zebra seem to disappear whereas the wildebeest stand out as dark forms.

Hunting in Groups. Another vexed question concerns the alleged setting of ambushes by carnivores that hunt in packs or family parties. Wild dogs, wolves and lions have been credited with springing ambushes. Lions especially have been described and diagrams given of their movements which show two or three moving round by a circuitous route to take up positions beyond a herd of antelopes. Then the rest of the pride spread out and close in on the prey, driving them towards these two. It is not uncommon to see one or more lions moving in on an antelope selected for pursuit, and to see the antelope take flight and run straight into the paws of another lion.

H. Kruuk and M. Turner, who have considerable experience of the behaviour of African animals, are of the opinion that the appearance of tactical

The chinkara, a race of the Dorcas gazelle in Pakistan. All gazelles are the target of land carnivores and they benefit from living in groups. A solitary individual has to rely on its own senses and resources for detecting the presence of a predator and the imminence of an attack. Living in groups the prey animals have the advantage of many pairs of eyes, ears and nostrils to give warning of danger.

A pack of Cape hunting dogs with their kill. They work as a team numbering usually 12 to 20 but sometimes as many as 50. They run down their prey by sheer endurance, their victims being typically large hoofed mammals like wildebeest, zebra, buffalo or gazelle.

co-operation is an illusion. They believe it is merely one animal taking advantage of prey put up by another predator. The same thing can happen when an antelope, startled by a motor-car, dashes for cover, straight into the waiting paws of a lion or leopard.

The evidence is contradictory and the truth may be that a simple form of co-operation may be possible and that anything more than this is due to over-enthusiastic interpretation by the human observer. There was a couple of dogs living in a house in southern England that made friends with the gardener, who went home each evening at the same time, on his bicycle. It happened many times that, just before the hour for the gardener's departure, the two dogs would be seen trotting down the lane the gardener would take, and at a certain point they would separate, each going into a ditch on either side of the lane. When, a few minutes later, the gardener on his bicycle reached this point, the dogs would simultaneously jump out on to him, barking. There seemed to be a degree of understanding between the dogs and there is no reason to suppose this would be impossible in two wild carnivores.

Nevertheless, there is need for caution in interpreting actions that look like ambushes, for there seems also to be misinterpretation in the repeated accounts of hunting dogs working in relays, those behind running forward when the leader tires. Hugo and Jane van Lawick-Goodall report that the pack always follows the leader. At times, however, some of the dogs in the rear may cut off corners so getting ahead. In that case the dog or dogs in the lead quicken their pace to regain the leading position. If incompletely observed, and for a few moments only, it could look like the rear dogs moving forward to take the lead.

Breeding Behaviour

No animal can live for ever. For this reason, to reproduce its own kind is one of the most important, if not the most important, functions that any animal can perform. There are two methods by which animals can reproduce themselves, the first involves a single individual and is called asexual reproduction, while the second involves a pair of animals, a male and a female, and is called sexual reproduction. The freshwater polyp *Hydra* is able to reproduce asexually by budding off young whenever conditions are right, that is, when there is enough food available to support new individuals.

Sexual reproduction is more complex. A cell from the male, called a sperm, fuses with a female egg cell, or ovum, to produce a cell called a zygote. This zygote grows into a new individual. Mammals employ the sexual method of reproduction only, and fusion of the male and female reproductive cells, or fertilization, takes place inside the female's body. The advantages of sexual reproduction is that, because individuals inherit genetic material from two parents, populations are produced composed of a great variety of individuals – individuals that may take advantage of new opportunities as they occur. The sexual habit does have some disadvantages however. A male mammal must find a mate of the same species but opposite sex, and typically he must introduce his sperm cells into the female at a time when she has produced ovum cells that are ready to be fertilized. There are exceptions, as we shall see later, in bats. It is not surprising therefore, that numerous mechanisms have evolved that make these things possible. Such mechanisms are often anatomical, for example males and females of one species are so built as to make the introduction of sperm into the female during copulation readily achieved. Other mechanisms are behavioural, and together, these are the subject of the present chapter.

Finding a Mate. In the great majority of mammals smell is the major sense and, as may be expected, scent is used to bring members of the sexes together. An acute sense of smell in mammals goes hand in hand with the possession of special scent glands. In many mammals they are a prominent feature, on the face, the feet or abdomen. Usually these are quiescent or working at a low level, except at moments of excitement. Some of these are used solely to bring male and female together. They become markedly active during the breeding season. Others are used for marking territories or for defence. A skunk, for example, uses the high potential of its scent glands under stress of alarm, as do other

◁ Donkey stallion displaying flehmen. This is a response on the part of the males of the hoofed animals to odours produced by the female, particularly of the urine when she is in oestrus. It looks like an expression of disgust and it is suspected that the turning back of the lips and the wrinkling of the nose serve to open the orifices leading into Jacobson's organ. The increased contact between the odours and Jacobson's organ is thought to have a sexually stimulating effect.

The White-tailed deer has a distribu- ▷ tion that extends from southern Canada to Peru. This superb male displays the antlers used in the ritual fighting that is a characteristic of the family to which it belongs.

members of the weasel family, and some do so also during the play that precedes copulation. Badgers give out a heavy smell of musk, for example. These odours are readily apparent to the feeble human nostrils. In other species there may be more subtle odours which we cannot appreciate. This may be why nobody can say what is the function of the gland on the snout of a capybara or the 'chestnuts' on the legs of horses, zebras and asses.

The voice is sometimes used to bring opposite sexes together, as in some dolphins which have special mating calls, and in a few land animals. Dogs call to each other, the male Raccoon dog has what is called a yearning call, foxes have a symbolic whine and the male Indian jackal has an entreaty call. These names merely represent the terms applied by the authors studying the different species. The female Black-backed jackal gives a mating call which is answered by the male.

Knowing when the Time is Ripe. Having found a possible mate, it is important for the male to be aware when the female has ova ready to be fertilized. Some mammals, for example man, the giraffe and elephants, are able to breed at any time during the year, but others have a particular breeding season. Female mammals undergo what is known as an oestrus cycle. At the start of each cycle the female is unwilling to mate but, as ova become ready for fertilization, she becomes more receptive and indicates this to any males that she meets. When she does this, she is said to be on heat, or in oestrus.

The word oestrus in Latin means a gadfly, an insect that goads and torments cattle. An alternative meaning in the original language is breeze. Both are appropriate to describe the female mammal coming into heat. A more dramatic and rhetorical account was given by a naturalist in Africa, many years ago. He described a group of eland bulls placidly feeding until a group of eland cows came into the vicinity. One or more of the cows was, presumably, in oestrus. The bulls 'were turned into a raging torrent'.

Scent is used by most mammals to indicate that they are in oestrus. This is well illustrated by the behaviour of the domestic dog. As the bitch begins to approach oestrus, the dog spends increasingly more time in her company. As the oestrus cycle advances, he may, from time to time, make preliminary moves to mate with her. She responds by moving away from him, giving a slight growl, which the dog readily accepts. The bitch may later stand in front of and at right angles to him, as if enticing him though without positioning herself for coition.

The effect is that the dog is held near the bitch until the peak of oestrus arrives and she is receptive. She has, however, become attractive to him in advance. So is achieved the end of all animal courtship: to ensure the presence of a male when the female has reached the point in her sexual cycle when fertilization, leading to pregnancy, is most likely to result.

The compelling force of a bitch at the height of oestrus can be seen when a dog, especially one that

During courtship, the Indian rhinoceros bull chases the cow and fighting often ensues. This eventually gives way to mating intention movements that culminate in mating, seen here, which may take as long as 80 minutes.

normally lives with the bitch, is separated from her at the crucial time because a mating is not desired by the owner. The dog is inconsolable. Food does not interest him. He spends his time whimpering, crying piteously and sniffing the air, and this despite the fact that the bitch may be locked up and a hundred yards or more away.

In many species of mammals the behaviour of the male is such that he is kept informed as to whether or not the female is approaching oestrus. The male kangaroo smells the cloaca and the pouch of any female he meets. In common with males of many other species he sniffs her urine when she micturates. Among these others are the antelopes, the males of which respond by a characteristic facial expression known as flehmen. This is a German word for which no English equivalent has been coined. On smelling the urine, especially of a female in oestrus, he raises his head, draws back his lips, wrinkles his nose and stops breathing for a moment. To the human observer it looks like an expression of disgust.

Smaller mammals seek their indicators of the nearness of their partners to breeding condition by nuzzling, sniffing or licking the genitalia. They may then urinate on themselves or their partners. Usually it is the males that perform these exercises but in a few species it is the prerogative of the female. The males of some species of deer adorn themselves to a marked extent, wallowing in mud, hooking up pieces of turf and throwing these adroitly onto their own backs and finally micturating on themselves.

Enurination – spraying urine on a partner – seems in some species to be more repellant, used by the female to repulse the male when she is in a non-receptive mood. The enurination by the male, it has been suggested, may be a means of asserting domination. It may, however, originate from the urination seen in some species, especially in juveniles, that expresses pleasurable excitement.

Courtship Ritual. In animals courtship is, in all but a few species of mammals, a preliminary to copulation, and it may be either very short or protracted. One of the briefest is in the elephant, in which the male lays his tusks along the female's

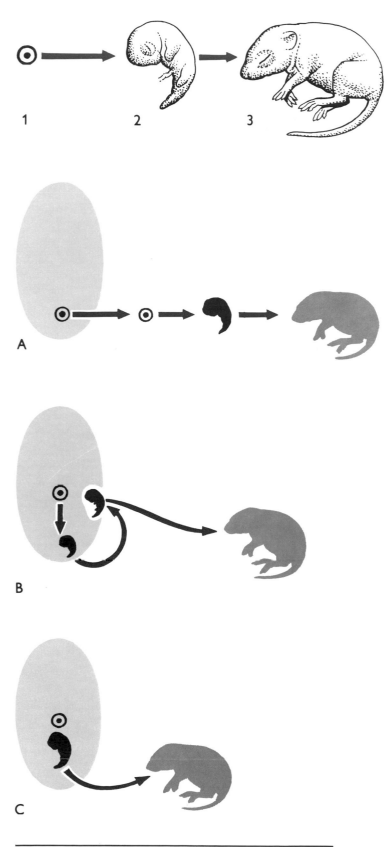

There are three stages in the early development of mammals, the egg (1) grows into an embryo (2), which then produces a young individual (3). Monotremes (A) lay eggs, and all three stages occur outside the mother's body. In Marsupials (B) the egg is retained inside the mother's body but the embryo is born in a very immature state and is transferred to a pouch. In Placental mammals (C), both the egg and the embryo are retained within the mother.

back as a signal that he is about to mount. If she is receptive, that is, if the moment is right for conception to occur, she stands her ground. If not, she walks away, and for the moment at least, that is the end of the affair. Rape, in the sense of enforced coupling with an unreceptive female is, with very few exceptions, found only in humans. It may possibly occur in monkeys and apes, but even this is doubtful.

In other mammals courtship rituals are often long and complex. In human courting couples there is a tendency towards what is called kitten play accompanied by a juvenile tone of voice or baby talk. When play forms part of the courtship, as in foxes, it follows precisely the play of cubs. Hamsters, squirrels and dormice, when courting, make sounds which resemble the distress calls of their young. The male Raccoon dog whines like one of his own whelps in distress.

In addition there may be crouching and posturing, and in some deer and many antelopes these may develop into ritualistic dancing and prancing. Much of this is probably no more than an expression of a growing excitement engendered in part at least by the receipt of odours from scent glands and urine.

The European hare, and to a lesser extent the rabbit, carries the body movements to excess, giving rise to the saying 'mad as a March hare'. The breeding season is prolonged, reaching a peak in March. The male hares seem then to go mad. With ears laid back, a male hare will buck like a horse, box with the front paws, bite, lash out with his powerful hind legs or leap straight up into the air in a series of writhing contortions. Often one will leap over another, lashing out with his hind paws and killing the other. When indulging in these antics the male hares seem oblivious of the human approach and one has been known to attack a dog – by accident.

Courtship in mammals is often accompanied by circling, as in the Roe deer, in which the male roe chases the doe around in circles, their hoofs wearing a groove in the ground, as he runs alongside her but slightly behind her, as if continually heading her off. Another typical example is that of the European hedgehog, a nocturnal animal and one that is strictly solitary. The male circles the female, at a distance of about a foot. She pivots to keep more or less face to face with him. So long as she is unreceptive she boxes him with her forepaws whenever he seeks to make a closer approach. Eventually the two mate and then go their separate ways. That the circling by the male may often continue for a long time before consummation, is evidenced by the circular groove trampled in the vegetation and soil. These well-worn circles in the ground are usually the only signs that tell of the mating of hedgehogs.

The male and female Grasshopper mouse, of North America, circle each other but this is only part of a more elaborate pattern of behaviour which may last up to three hours. During this time the courtship may be broken off and resumed several times. The circling is interrupted by the two rising on their hind legs to bring noses together in a 'kiss' or for the female to drop onto all fours to sniff his genitalia. The performance is concluded, when the female is fully receptive, by the male parading backwards and forwards in front of her, after which he crouches to pass under her belly, rubbing his back against it. Then he rises on his hind feet and grooms her face and neck. She responds by moving her tail to one side and raising her hindquarters.

Copulation may be brief, a matter of a few seconds, as in most cloven-hoofed animals, or it may last several hours. In the polecat, the pre-mating preliminaries are accompanied by the male biting the female's neck and drawing blood, and

Mating between lions. In the left picture the male prepares to mount while, during mating, the male seizes the female's neck in a way similar to that seen in tigers and other cats.

In the langurs shown here the female is in oestrus and is presenting to the male (left). That is, she turns and directs her hindquarters towards him, shakes her head and lashes her tail and invites copulation.

copulation lasts up to three hours. In the related sable it may last 18 hours. In some species it may be brief but repeated many times in a few hours as in Shaw's jird, a desert rodent of Tunisia, for which 224 copulations in two hours have been recorded. In the majority of mammals coupling is effected by the male mounting the female, known as dorso-ventral. Belly-to-belly coupling is reported for Two-toed sloths, some hamsters and few others such as whales and man. Inquisitive laymen sometimes ask how hedgehogs and porcupines mate. The answer is that they do so in the normal way, the female relaxing spines or quills, the male standing on hind feet and resting his forepaws lightly on her prickly armature.

Ritual Fighting. It has often been stated that male mammals fight one another for possession of available females. Male Red deer for example, round up females during the rutting season and then spend much energy running around the perimeter of the assembled harem, presumably both to prevent the females from straying and to keep watch for intruding males. Fights, conspicuous for the sound of clashing antlers, are not uncommon. It appears, however, that the male is defending territory rather than ensuring possession of the females. Indeed, it happens not infrequently that while the occupying stag is fighting an intruder, one or more other males will enter the harem and couple with females.

In the antelope known as the Uganda kob, the males mark out territories and fight to retain them. The females enter the territories voluntarily, so in effect choose their own mates. Any fighting between males does nothing to alter this. It only alters the boundary of a territory.

Once this became clear, it then became apparent, in species after species, that territory is the fundamental cause of fighting between males. The fact that so often a female is nearby, lends colour to an

61

appearance of the fight being for possession of her. There may be a few species in which this does in fact happen, but they need more critical examination than has hitherto been carried out before this can be stated with certainty.

Birth. The mammals are classified into three groups each one characterized by a different method of breeding: the monotremes lay eggs; the young of marsupials are born in a very immature state and spend the early part of their lives in a pouch on their mother's belly; while in the placentals, the young are retained inside the mother for a much longer period and are relatively well developed when born. To understand why these differences occur, we must realize that the mammals have evolved from a group of reptiles which lived some 200 million years ago, and that these reptiles were oviparous that is, they laid eggs. The monotremes are the survivors of a group of mammals that retained the reptilian method of reproduction, while the marsupials and placentals are groups in which new methods of breeding have evolved.

Monotremes. The monotremes, confined today to Australia and New Guinea, include only a few species of echidna, or Spiny anteater, and the platypus or duckbill. Their eggs are soft-shelled, the shell being parchment-like. The female platypus retires to a burrow to lay her two eggs. When, after only 14 days, these hatch, the babies are nourished with milk that oozes from glands on the mother's abdomen, there being no true teats as in placental mammals. The female echidna has a pouch on the abdomen to which the single egg is transferred, the manner in which this is done being still largely a matter of speculation. The young, on hatching, are fed in a manner similar to that of the platypus.

Marsupials. The marsupials, or pouch-bearers, include the opossum, Tasmanian wolf, Koala bear, kangaroo and wallaby. They are now confined to Australasia and South America, with the exception of a single species, the Virginian opossum, that occurs in North America. The method by which these curious animals reproduce themselves was unknown until quite recently. A Dutch sea captain called Pelsart had seen young in the pouch of a wallaby, and had concluded that they were produced by budding from the teats. Later explanations were that the mother lifted the newly-born young with her fore paws and placed it in the pouch, lifting it with her lips to do so. In 1830 and again in 1882 the truth of marsupial birth was revealed by amateur naturalists but ignored by scientists. It was not until 1923 that W. T. Hornaday, then director of the New York Zoological Gardens, confirmed it.

To illustrate reproduction in marsupials, let us trace the events in the early life of a kangaroo. The female produces a single ovum every four to six weeks depending upon the species. If copulation takes place successfully, an egg is fertilized, that is, it fuses with a male sperm cell. The egg, now called

◁ Red deer stags sparring with their antlers. In the course of this the two males assess the measure of each other's strength. Accidents are infrequent because the tines, the points on the antlers, are not used directly as offensive weapons.

The Spiny anteater of Australia, one of ▷ the two types of egg-laying mammals. It is still a mystery how the female manages to transfer or deposit her egg in the pouch which appears on her abdomen at the start of the breeding season. It seems that there is a period of gestation in the uterus lasting 27 days before the egg is actually laid. When the young hatches it is very tiny and clings to one of two milk patches on the mother's abdomen. Spiny anteaters do not have teats. The young Spiny anteater remains in the pouch for an unknown period until it starts to grow spines. The mother then places it in a burrow, and visits it every two days to feed it until it is weaned three months later.

a zygote, is retained inside the mother and begins to divide. After 29 to 38 days the embryonic kangaroo is ready to be born. During the period of pregnancy, the gestation period, the embryo receives oxygen and food from its mother through an organ called the placenta.

Just prior to the birth the female kangaroo cleans the inside of her pouch with her tongue, holding the mouth of the pouch open with her forepaws. She then rests with her back against a support with her hindlegs and tail stretched out in front. At birth the young kangaroo is less than an inch long, hairless, blind and deaf, the inner ear being incomplete. In order to complete its development, the infant must find its way to the mother's pouch, and it does this by crawling through her fur guided by its sense of smell. This is an extremely difficult journey for such a tiny individual to embark upon, but it is made easier by the fact that its forelegs, with which it grips its mother's fur, are relatively well developed at birth, and are longer and stronger than the hindlegs. Once it has reached the safety of the pouch, the young kangaroo takes one of four teats into its circular mouth and remains attached to it

The birth and growing up of a wallaby is a most extraordinary process. First there is the strange method of pouch birth with the young being born in a very undeveloped state. Then follows the almost unbelievable enlargement of the pouch to accommodate the young Joey. Finally, there are the skilful gymnastics required of both mother and offspring in the later stages of their association by which the Joey can re-enter the pouch by leaping into it headfirst.

During mating, the male tiger seizes the female's neck with his teeth. This behaviour is a characteristic of all cats.

for at least 40 days. During this time it feeds on milk produced by the mother using its strong muscular tongue for sucking. After 190 days, the kangaroo is ready to make its first excursions to the outside world.

A female kangaroo can mate soon after having given birth. As a result, an embryo sometimes develops in the mother while her pouch is still occupied. To avoid overcrowding of the pouch, a process called embryonic disapause comes into operation. Embryonic disapause simply means that development of an embryo is arrested at an early

stage until the pouch is vacated and ready for reoccupation.

Placentals. The placentals include all the other mammals living today. Their great success is due in the main to their efficient method of reproduction. The young are retained inside the mother for a much longer gestation period than in marsupials, and are as a result efficiently protected during the most vulnerable period of their lives.

The length of the gestation period is related to two factors, the size of the animal, and the state of maturity of the young when born. The shortest

During the period of heat the male impala licks the genitalia of the female, rises to his hindlegs and mounts. After copulation he becomes aggressive and chases the female.

recorded time for placentals is 15 days for the Golden hamster, and 760 days—over two years!—for the Indian elephant. For mouse-sized placentals gestation lasts around 20–30 days, for fox-sized placentals it is about 50–60, for pig-sized about 140–150, and for most of the larger hoofed animals about a year.

In some mammals, the young are completely helpless when born and pass their early days in a nest. Such young require careful nursing by their mothers and are called altricial young, after the latin word *altrix* meaning nourisher. Precocial young, on the other hand, are born in a more advanced state and are able to stand, even to walk, unaided within an hour. In general, the chances of survival are less for altricial young than for precocial young. Therefore litters of altricial mammals tend to be large while those of the latter are small.

A fox cub is typical of altricial young. It spends 51–63 days in the womb and is born into a nest of hair that the vixen pulls from her front with her teeth. This exposes the teats. The cub is born helpless, almost naked, blind and deaf. Eyes and ears open after about 10 days. The mother curls herself protectively around her litter enabling them to find her teats with the minimum of trouble. The cubs remain below ground for 3–4 weeks and even

when they come above ground their limbs are not fully co-ordinated, and they tend to totter as they walk.

Baby mammals born above ground on open ground must be able to move in a short while. Enemies are either waiting nearby or will soon be on the scene, and to survive they cannot afford to rest for more than a short while before bringing their limbs into working condition. In contrast to a fox, which has babies lying down, precocial mammals such as horses, cows or sheep, stand to deliver their offspring. As in all mammals, the newly born is licked clean of the birth membranes while the placenta is eaten by the mother, whether she is a herbivore or a carnivore. For a while the baby lies resting, its eyes fully open, its ears already functioning and its body fully clothed with hair. Then it begins to struggle, feebly at first, then more vigorously, its aim being to get onto its feet, which it finally manages with a determined heave. So it stands, unsteadily at first, tottering as it takes its first step and looking all the time as if its legs are about to give way beneath it. Probably it falls, collapsing to the ground, but after a brief pause to gather its strength it tries again. This baby has been in the womb for nearly a year, in contrast to the

House mice are found all over the world. The secret of their success lies in their reproductive capacity which varies according to the habitat. For example, where cereal crops are stored in ricks, a female mouse may have ten litters per year.

two months of the fox cub, but it has been born at a more advanced stage of development and, within little more than an hour after birth, it is able to run beside its mother.

Retention of the embryo inside the pregnant mother is, as noted above, a method of breeding that affords the maximum amount of protection for young mammals. There are however, disadvantages, the chief one being that a single female produces fewer offspring than would otherwise be possible. It is not surprising that, in a number of placentals mechanisms have evolved that make both retention

of young and a high birthrate possible. In the European hare for example, super-foetation may occur, the female may mate while pregnant and so ends up carrying two sets of embryos of different ages.

In most placentals, pregnancy and birth follow fertilization without any obvious delay, but there are exceptions to this general rule. Small insect-eating bats that live in temperate latitudes hibernate during the winter. Often they mate in the autumn rather than during the spring. Obviously, it would be inconvenient to break the period of hibernation

The crowded interior of the pouch of a female opossum where the babies are firmly attached to the nipples and are already in a fairly advanced stage of development. They leave the pouch when 11 weeks old.

in order to give birth to young, and so the breeding process is delayed. The female bat stores male sperm cells in her body until the spring, and fertilization takes place only after she has emerged from her hibernation. Birth then occurs 35–45 days later. Similar mechanisms are not found in tropical bats because under constant climatic conditions breeding is possible throughout the year.

Number of Offspring. The size of the litter in mammals varies considerably. A Polar bear normally has 1–4 cubs at a time, while, at the other extreme a tenrec, an insectivore from Madagascar that resembles a hedgehog, was once found to contain 32 embryos. The reason for such variation is that the young of different mammals have different chances of survival. Normally, the rate at which offspring are produced equals the mortality rate of that particular species, so that the population remains more or less constant over long periods. Such population maintenance is illustrated by the rabbits on the island of Skokholm off the coast of Wales. The island is uninhabited except for the keepers of the lighthouse. There are no predators. Each year the population of wild rabbits is about 2,000. After the breeding season it rises to around 20,000. By the beginning of the following year it has been reduced to 2,000. Only sufficient of the young survive to replace adults dying during that period. The rest die of accident, congenital defects, neglect on the part of the mother, disease, malnutrition or any one of a dozen or more natural hazards.

Maternal Care

Parental care reaches a higher level in mammals than in any other group of animals. Live birth means a heavier burden for the female and prolonged post-natal care of the young imposes strains, hazards and responsibilities not found elsewhere in the animal kingdom. Yet so strong is the hormonal action underlying the maternal instinct that the females do not shrink from these. Indeed, there is often an urge to add to them unnecessarily, such is the bond between female and infant. It is by no means unknown for the female to adopt the young of another parent, even one of another species. She may even steal it. This occurs when a female has lost her own offspring and it may take place also towards the end of a pregnancy or when a false pregnancy occurs. When this happens in the human species the unfortunate individual receives little sympathy but is treated as a felon.

With a few exceptions, the phenomenon spoken of as parental care ought with justice to be called maternal care, since it is the female who bears the brunt, and usually the total burden of rearing the young. In many instances they are with her for a few weeks only, in others the protective period may last for months. In a few instances it may be years. As with gestation, so with maternal care, the period tends to be related to the size of the animal. In the smallest mammals, the smaller shrews, the young are weaned at a little over three weeks and become independent shortly after this. Young elephants are weaned at two years and may stay with the mother for some time after that.

Crèches and Caravans. Nobody pretends this is an inviolable rule and there is one notable exception, the giraffe. When it is only a few weeks old a young giraffe leaves its mother and joins other juveniles of about their own age under the surveillance of two adult guardians. Should one of this group of youngsters wander away, or become separated, when playing, for example, one of the guardians will round it up and restore it to the group. The young giraffe also starts to eat leaves and twigs when a few weeks old although it will continue to suckle for six months or more. It will also rejoin the mother for a while at intervals, only to return to the 'crèche', and it will not finally leave her until it is a year or more old.

Hippopotamuses have a similar system although theirs is more like 'baby-sitting'. In a colony the females with infants and juveniles of both sexes

The infant chimpanzee is carried on its mother's belly at first, clinging tightly, a reflex well-developed from birth. Sometimes the mother puts her baby on her back, feeding it when it asks for food. While she is feeding or building her nest for the night she may put her infant on a branch beside her. Even if it screams with temper it is generally ignored: indeed the whole chimpanzee band carries on with its activities as if the youngster were not there. Even after the young have become independent they still associate with their mothers up to a year after the birth of a second infant. At this stage although the juveniles still build their nests near the mother's they associate more with other youngsters, wrestling, chasing and swinging about in the trees.

occupy the centre of the territory. The mature males are solitary, each holding a territory on the periphery. When a mother hippo goes on land to feed she may leave her infant in the care of one or more females, in the company of other infants forming the crèche. There is also purposive education of the young, for its greater security.

The main enemy of infant and juvenile hippos are the adult males who will attack them whenever they have the opportunity. To protect her youngster the female hippo teaches it to swim at her shoulder, which is the best position in which she can defend it in water. When it accompanies her on land it must walk behind her. Should it fail to do so she chastises it by beating it with her huge muzzle until it 'gets the idea' what is required of it.

The pattern of behaviour, for both giraffes and hippopotamuses, is elastic and tends to vary according to whether the herd is small or large. Some sharing of responsibility is clearly needed, especially by female hippos, since the gestation period is eight months and the baby suckles for a year.

In large mammals single births are the rule, twins being rare, and often these do not survive. In small animals, constantly under pressure from predators, large and frequent litters are needed to maintain population numbers. Large families are less easy to control and crèche systems or baby-sitting would be impracticable. Nevertheless, there are ingenious protective methods to be seen, the most remarkable being that of one of the White-toothed shrews of Europe. A similar behaviour is reported for the Musk-shrew of south-east Asia, which also forms 'caravans'.

When the female White-toothed shrew takes her family from the nest, to find food for herself, one of the babies grips her tail near the base with its teeth. The next baby grips the tail of the first and so on until the whole family of half-a-dozen forms a long line with mother in front and the babies hanging onto each other's tails. To be effective all must move in step, halt when mother halts and step off again when mother starts to move. A family of caravanning shrews running rapidly is sometimes mistaken for a snake. When the family returns to the nest the caravan breaks up. This behaviour persists for the first three weeks after birth and is only possible because in White-toothed shrews the young are precocious as compared with the young of other shrews. The caravan holds together so firmly that the whole family can be lifted into the air by lifting the mother.

Caravanning is a two-way process. The babies instinctively form up and grasp the tail in front. Should any fail to do so or be slow in doing so the mother waits for them. Also, if the chain is broken she will wait for it to reform before moving off once more.

When baby hedgehogs are able to walk the mother takes them with her when she goes out on feeding expeditions. The family commonly travels in single file, the mother in front, the youngsters nose-to-tail behind her. Weasel families not uncommonly do the same, and in principle it is the same as the female hippopotamus with her one youngster following behind. In the smaller mammals the behaviour may be instinctive and automatic, without the need for teaching the youngster by chastisement. There is, however, one reputable record of a Water vole swimming with a young one behind her. The youngster strayed off course, there was a flurry of water as the mother turned and appeared to be chastising it, after which it followed its mother obediently.

The caravanning of White-toothed shrews also recalls another form of family travel. The female Wood mouse, or Longtailed field mouse, of Europe, takes her babies on expeditions clinging to her teats. The babies are not in single file but running abreast. The whole family moves remarkably quickly so that it is difficult to see precisely the details, but they appear to be running in step. Sometimes the mother will jump an obstacle. The babies maintain their hold and appear to suffer no harm – certainly there is no upset to the rhythm of their running – from the shock of being pulled through the air and then landing. The same behaviour has been observed in the Swamp rat of Africa and the Deer mouse of North America. It must have considerable survival value judging by the abundance of both the Wood mouse and Deer mouse.

These caravans and semi-caravans are in contrast to the usual method of transportation of the young, which is by carrying them one at a time in the mouth. This is a familiar procedure for the domestic cat and dog and is used widely by other carnivores as well as by rats, mice, squirrels and others.

A pregnant female Polar bear goes into a state of partial ▷ hibernation, known as winter dormancy, at the onset of winter. In her hideout in the snow her one to four cubs are born in January or February. They remain with her until the following autumn. The cubs are very small, weighing no more than 1½ lb (0·68 kg) at first, but they make great demands on the mother and grow rapidly.

Anteaters of South America and pangolins of the tropics of the Old World feed on ants and termites and have very small tubular mouths, useless for carrying babies. Typically they have one young at a birth and this, born in a precocious stage of development, is carried about by the mother clinging to her tail. Sloths, tree dwellers of tropical America, carry their babies clinging to their underside, much as monkey babies are carried.

Even when the infant sloth has passed the stage of being carried the mother is still considerate for it, as it follows her through the trees. If a gap in the branches has to be crossed she will draw the branches together and hanging between allows the youngster to use her back as a bridge. Unfortunately there are too few careful observations on this aspect of maternal care. From the few scattered records we may suspect that education in living is a continuous part of this transport. For example, a baby Slow loris clinging to its mother's back as she moved through the branches was compelled to climb over a branch as the mother squeezed under it, resuming its hold on the mother after negotiating the branch.

The more usual form of this kind of transportation is that typical for monkeys, in which the baby clings to the mother's chest or belly, holding onto the hair with its grasping hands. There may be times when the mother assists by putting an arm round her baby but the more common action is to use the arm to adjust the baby to a more secure or comfortable position.

Hardly more comfortable for the mother with a baby grasping handfulls of hair must be the device found in bats. The females carry their babies around, even when out hunting insects on the wing, until they are too large to be carried. The single baby – twins are not uncommon in some species – holds on by a false teat. The two functional teats are on the chest but there are two non-functional or false teats in the groin, one of which the baby holds in its mouth when not feeding. The grip on the teat is assisted by special hooked milk teeth in the front of the baby's jaws. The male bats take no part in rearing the young. The pregnant females associate in breeding roosts apart from the males.

This extra weight for a flying animal must constitute a near intolerable burden when there are twins, and mother bats have to leave their babies behind, hanging in the roost, when they grow too big to be carried. This results in a high mortality among the young, from losing their grip and falling to the floor. There they soon succumb to the ammonia-charged atmosphere, from the layer of guano that characterises every roost. The females make no attempt at rescue, apparently, although if a baby falls to the ground elsewhere and utters its high-pitched distress call, the mother, and other bats, even of other species, will swoop down as if

A female Kerguelen fur seal and her baby on the rocky shore preferred by these animals. The babies are born in November and December. At birth they are 1½ ft (0·46 m) long, clothed in a black woolly fur which is moulted during April giving way to the grey yearling coat. About this time they are weaned.

The Pacific striped dolphin found in the northern Pacific lives in large socially organized schools. The young dolphin, like the hoofed land animals to which dolphins are closely related, must be able to move independently from birth. The mother may assist her baby, especially in the few moments after birth, by raising it to the surface to take its first breath.

bent on a rescue attempt. Most observers claim that bats are unable to retrieve the young but there is at least one authentic observation of a baby bat on the ground being rescued. An adult flew down, spread itself over the baby and when it flew away the baby was no longer to be seen, and was presumably hanging on to the adult.

The extent of the burden on the mother can be gauged from recent observations on laboratory mice. In these it was found that the larger the litter the more time the mother spent out of the nest. With a small litter she spent much time in the nest, her foraging expeditions were brief. With large litters the mothers spent significantly more time feeding, a natural consequence of having more mouths to feed. She also spent more time outside the nest resting and grooming, and it is assumed that in a well-filled nest she suffers from disturbance from her restless litter, cramped conditions and over-stimulation of the mammary glands.

The Role of the Male. Some rodents show another form of solicitude for their young. The behaviour pattern involved is undoubtedly innate yet its working suggests the sort of unselfishness which is normally assumed to be the prerogative of thinking man. In these species, which include the Wood rat and the Prairie dog, both of North America, male and female care for the litter. The Wood rat builds a 'house', a nest of sticks with several chambers, the whole roofed with sticks to keep out the weather. When the litter is becoming independent the parents in both these species leave them in occupation of the nest and go elsewhere to build a new nest for themselves.

Because, in these two instances, the male also participates in this unselfish behaviour, we are reminded that sometimes the paternal parent does share responsibility. It may be only in defence as when a herd of Musk-ox, of Arctic regions, form a protective phalanx against wolves. The males take

73

the leading role in this, one after the other stepping forward to do battle, their places being taken by others when they fall. These tactics may be successful against wolves but not against men armed with rifles. There have been instances in which a Musk-ox has been shot for its pelt, for use as a sleigh rug and, so readily have the rest come to its aid, the whole herd has had to be shot. The same has happened when a calf has been wanted for a zoo. All the adults have had to be shot to secure one juvenile. That is why the Musk-ox numbers dropped dramatically since the animal was first seen by a European at the end of the 17th century. The species is now protected.

No more striking examples can be found of paternal participation than in the European Red fox. Throughout the non-breeding season this fox is solitary but in December, pairs begin to be formed. Little is known of the pre-mating behaviour in the wild but in captive foxes the male is greed personified until the cubs are born. He will eat all the food if precautions are not taken to feed dog-fox and vixen separately. The moment the cubs are born this changes radically. Then, the dog-fox takes every scrap of food into his maw, takes it to the entrance to the earth where the vixen is underground with her cubs and calls to her. When she appears at the entrance he drops all the food in front of her and eats nothing until she is replete. This continues until the cubs are approaching partial independence.

Something of this sort must happen in the wild, also. Strangely, it has long been asserted that the dog-fox takes no part in rearing the cubs. This is remarkable because the male can often be seen with its maw full of food heading for the earth where a vixen and cubs are known to be. Moreover, by watching at an earth about sundown the two parents and the cubs can be seen coming out together, when the cubs are well-grown, pausing to play for a while on a patch of greensward, then setting off together, presumably for hunting. To clinch the matter, instances are known in which a vixen and her cubs have been dug out and killed. Later, it has been found that one cub was overlooked and that the dog-fox, returning after the massacre, has continued to feed and rear this cub.

It may well be that more examples will accumulate in time, of paternal devotion. In the meantime, more stress tends to be laid on instances in which the mother has to defend her offspring against their father. Examples of this have been recorded for animals as diverse as hippopotamuses, wild cats, domestic cats and several species of rodents. The evidence, scattered and scanty as it is, suggests that in species in which the father is normally guilty of potential infanticide there are occasionally model fathers. Conversely, in species, such as the Red fox, where he is usually a solicitous parent, there may be fathers that neglect or even attack the young. Such anomalies are not wholly absent in the human species.

Animal Midwives. It is tempting, if not inevitable, in discussing parental care in animals to draw attention to similarities in the behaviour of humans, the species in which this care is most organized and prolonged. After baby-sitting hippopotamuses and good and bad fathers, we come to what is close to midwifery. The two best examples are found in elephants and in dolphins and porpoises.

Those whose business has taken them into the bush in Africa have several times recorded what happens – or appears to happen – when a cow elephant gives birth. A typical example tells of a female elephant, obviously approaching the end of pregnancy, disappearing into a thicket with two other females in attendance. What took place there could not be observed. All that could be recorded for certain was that some time later the three females re-appeared accompanied by a newborn calf. It is surmised that either the accompanying females were on hand to assist the birth, to support the mother by standing on either side of her, or to be on hand to help drive off a would-be predator.

The second of these may be the more likely. There have accumulated over the years several instances of elephants seeking to rescue one of their number that has been shot or injured. They do this by helping it to its feet and then ranging themselves on either side to keep it from falling and to assist its locomotion.

The comparable behaviour in porpoises and dolphins was wholly unsuspected until the first seaquarium was built in Florida, about 1939–40, and for obvious reasons. A large animal living wholly in deep water evades observation. In the seaquarium it was found that the young dolphin or porpoise is born tail first and as soon as it is clear of the birth canal it rises to the surface to take its first breath. The mother snaps the umbilical cord by a quick about-turn, which also brings her in the best

Tree porcupines, as their name suggests, live almost ▷ entirely in trees. The young are therefore vulnerable and require protection from the mother during their early stages of development.

position to assist the newborn calf to the surface, lifting it up by putting her snout under it. She may even hold one of its flippers in her mouth to assist its upward journey.

All this takes place with two other females in attendance, one either side of the mother. The reason for this seems obvious. Sharks, with their highly developed sense of smell, readily home on blood or bare flesh, including carrion. Anything injured or bleeding is likely to be attacked. Blood lost into the water during parturition, as well as the placenta, when it is expelled, could attract sharks which, once on the spot might attack the mother or the calf. Porpoises and dolphins are a match for sharks provided they have no other matters to attend to and are not fighting single-handed.

The devotion of some animal mothers to a still-born baby, or to one that dies soon after birth is pathetic to see. A female monkey has been seen to carry a dead baby about until it decomposed. Female dolphins and porpoises in similar circumstances try repeatedly to resuscitate their dead babies by lifting them again and again to the surface to breathe.

This contrasts strongly with the attitude of seal mothers, and even more of the fathers, to their offspring. Although the detailed pattern varies from species to species, the general plan of breeding in seals is of a dominant bull with a harem of cows. The bull reaches the breeding beach first, followed by a number of cows who shortly afterwards give birth. Mating then takes place with little ceremony or preliminaries. All this time the bachelor seals, the young males, are challenging the occupying bull. Fights take place at any point on the boundary of the territory the dominant bull rushing over to meet and drive off the challenger. Any pup in his path is exposed to the danger of being crushed. The bulls do nothing to avoid injuring the pups and the cows are powerless to prevent him.

Care of the Young. An important element in parental care, and particularly in maternal care, is the licking. This starts with the birth of the infant and is, with some exceptions, general throughout the order Mammalia. It has been described in some detail for the kangaroo and this description is applicable to most species. As soon as parturition is achieved the mother starts to lick the newborn infant .to clean its coat of birth membranes and fluids. It is also the first step to forging a bond between parent and child.

Licking continues throughout the period of infancy. Its main purpose is probably that of surface hygene but it has side-effects. Experience in hand-rearing baby mammals suggests that it may have a therapeutic value. For example, licking the baby's abdomen provides a massage keeping the digestive organs in good working condition. In hand-reared baby mammals a good substitute, when digestive disorders have occurred, is provided by massaging the abdomen with a soft brush. It is of interest to note that minor digestive disorders in humans of all ages can be eased or cured by gentle massage with the hand or a soft brush.

Young Japanese monkey *Macaca fuscata* with its mother feeding on the shore. This species supplies good evidence that young mammals, like children, learn a great deal by imitation of their parents and also from companions of their own age. Japanese monkeys were fed by scientists with sweet potatoes on a sandy shore of an island. There was no particular intent to observe peculiarities of feeding habits so far as the Japanese scientists were concerned, but they did note fortuitously that one baby monkey took each potato into the sea to wash the grit off it. The habit spread by imitation to the other young monkeys and then to the adults.

The angwantibo of the tropical forests of West Africa is a primate that stands halfway between the lemurs and the monkeys. Most of the lemurs have their litters in a nest. The angwantibo and its relatives, the lorises, use the method of transporting their baby favoured by monkeys. There is only one baby at a birth, after a gestation of about four months, and this climbs directly on to the mother's fur, spending most of its infant life riding pick-a-back.

In many species of mammals, notably the carnivores, the nursing mother swallows the excreta of the infants, so keeping the nest clean, possibly also assisting concealment by obviating any risk of a strong tell-tale odour. Early in the life of the baby the mother licks the anus, which seems to have the effect of assisting the evacuation of the bowel. It is this that probably leads on to the swallowing of the excrement.

Experiments with laboratory rats have shown that babies that are handled daily are more hardy, lively and intelligent than control litters kept under precisely the same conditions of comfort and feeding but never handled. Being handled can be interpreted as receiving gentle surface massage and is probably the equivalent of being licked. It seems reasonable to argue from this that small litters have the additional advantages of receiving more licking, and therefore have an additional chance of survival.

Parental care includes, in altricial species at least, the provision of a shelter and usually a soft bed, of dry vegetable matter such as grass or leaves, supplemented in many species with hair torn from the maternal body with the teeth. It also includes defence against predators, in which is seen further evidence of intensive, even heroic, maternal devotion. In precocial species the young has the instinct to 'freeze' at the first signs of danger, its coat often camouflaged as in the dappled coat of many deer fawns, so aiding concealment. The mother takes refuge in flight, tending to draw the attackers away from the concealed infant, or else stands and fights, according to circumstances.

Motherhood tends to increase courage, in defence of the young, even to the point of the mother fighting a predator from which at other times she would flee. This heightened maternal courage is epitomized in the story of the doe rabbit with young that lashed out with her hind feet at a stoat, traditional enemy of the rabbit, sending it through the air a distance of 16 ft (4·9 m) or more. The stoat did not return.

Parental care, especially in carnivores, means providing food for the offspring when the end of weaning is near and they are beginning to chew solid food. For the most part this entails no more than extra hunting, to provide food for both parent and family. And even when the male assists in this it does not wholly relieve the mother of extra work. In providing food, also, deeds of near-heroism have been recorded. This is epitomized in the account of the elderly, very emaciated lioness with damaged claws, unable to pull down even small game. She was seen trying to tear apart a dried hide to feed her cubs.

Play. Finally, in the more advanced species, in terms of mental equipment – and this applies especially to the young carnivores – there is the need to teach them to hunt. This is probably the lightest burden of all because to a large extent it is assisted by the young animals play.

Whether it is justified to speak of play in animals has been contested in the past. The origins of play have also been the subject of disagreement. Today, ethologists, students of behaviour, seem not so averse as they were a few years ago to using the word for juvenile activities. And whatever may be the origin of play activities, whether due to an excess of energy, according to Herbert Spencer and Schiller, or as a preparation for more serious activities of adult life, there is good reason to suppose that the play pattern is innate, or what would be popularly called instinctive.

Another explanation of play provided by the philosophers is that it is an activity due to freedom from over-riding care. This is evidenced by the observable fact that play is more frequent and prolonged in the young of carnivores, whose parents by their powerful claws and teeth can afford the greatest security. It cannot be denied also that this is correlated with the greater need in carnivores of physical agility, quick reactions and well-co-ordinated muscle action in pursuit of prey.

Play, in its simplest form, in deer fawns and in calves, consists of little more than chasing each other, with little or no participation by the parents. In these herbivores speed in flight is the most important single means of self-protection in the adult. Nevertheless, more careful observation has disclosed other components in play, including actions resembling coition even before the sex-organs have matured.

In the related sheep and goat the outstanding features of play are leaping and also jumping onto any prominence, often no more than the mother's back as she rests crouched on the ground. Both sheep and goats are by nature mountain animals for which the ability to leap from rock to rock is essential. In the play of the young sheep or goat,

The young giraffe may be born at any time of the year after ▷ a gestation of 14½ months. At birth, it is about 6 ft (1·8 m) tall. It can stand and suck within an hour, and walk or run soon after.

support is given to the view that play is a preparation for the more earnest activities to come. Moreover, this innate pattern still survives in domesticated sheep and goats that for generations have lived in lowland pastures.

It is possible to write about play in animals only in general terms since there are so few detailed observations on it. The play of kittens and puppies is familiar to most people and can act as a standard for most carnivores. In kittens it consists of one stalking the litter-mate on flattened belly with tail lashing, leaping into the air as if clawing at a flying object, throwing small objects into the air and pouncing on them as they fall to the ground and chasing or stalking moving objects, such as a ball. Puppies tend to do more chasing and retrieving, for example, of a ball or a thrown stick, so emphasising the essential difference in the hunting methods of the cats and dogs. Typically, cats stalk their prey and kill with a final pounce and dogs run their prey down, killing them with slashing bites.

One fact readily emerges from watching the play of kittens and puppies: that the pattern is constant for each species. This leaves no doubt it is innate, or instinctive. Some of it may be varied by learning but mainly the pattern is constant for any given species. Nevertheless, there is no reason to suppose it is not

enjoyable or that fun and pleasure are not components of it.

Traditionally, adult dolphins and porpoises play. The whole school joins in and for a while indulge in a mixture of aquabatics, standing on their heads, swimming upside-down, leaping from the water and swimming in line abreast.

Since seaquaria came into being, serious study of their play has disclosed that dolphins and porpoises not only initiate games with their attendants they even invent fresh games, and there is a strong suspicion amounting to a virtual certainty that these animals 'play for fun'. This may be correlated with the greater cerebration of this type of animal but it does nothing to rule out the possibility that other animals play for fun, or at least derive satisfaction from their play activities.

One animal in which play has been studied in detail is the European red fox. A cub, hand-reared from before the time its eyes opened, was kept in a pen furnished with branches and logs to simulate a natural habitat. When it first started to play there were three outstanding features, apart from running around rapidly and seizing objects with its teeth and shaking them vigorously. These were: leaping about on all fours, suddenly flopping to the ground with its chin on its paws, and driving its nose into

◁ **Two aspects of maternal care are exhibited here by a mare and her foal. She is suckling her young foal and, at the same time, grooms it.**

Many mammals change the site of the nursery nest ▷ several times before their youngsters are old enough to be independent. This may be because there has been disturbance near where the youngsters are living or there may be other reasons for it. The usual method of transferring the young from one place to another is for the mother to take the baby in her mouth holding it either by the loose skin of the neck or by the belly or even by taking its head inside her mouth. In all cases the baby is held delicately and gently and is carried without injuring it, even by the big cats with their powerful teeth.

The Ringtail lemur lives in troops of up to 24 members which move about in a fairly well defined home range. The young travel for the first two weeks of their lives clinging to the underside of the mother, but later, they ride on her back as seen in this picture.

the ground. These actions are clearly of great value to a fox, especially the sudden flopping to the ground in the middle of its running. Even in short grass the animal seems to disappear. Pushing its nose into the ground later evolves into the search for mice and voles.

An important aspect of the play of a fox cub is that its pattern changes at the end of the first fortnight. Then a new set of play actions is seen, with the old actions being included from time to time, as if the cub were refreshing its memory of them. This second stage of play also lasts about a fortnight, with a third stage of about two week's duration in which a third set of play actions is used interspersed with those used in the first two stages.

In the following year, the fox, now mature, was given a mate who produced a litter of cubs. The play of these cubs was identical with those of their father, when he was at their age. Moreover, many of the actions used by the cubs when playing together were identical with some of those used by the parents in the pre-mating behaviour, or courtship.

Territory

The majority of mammals spend most of their time within a well-defined area of land or water. Each in its own area is thoroughly familiar with the topography of it and can make its way around it speedily and with confidence. It knows the best places to find food, shelter, bolt-holes when pursued, and drinking places. Within this area it has everything it needs, including a sense of security. This is its home range. Somewhere near the centre of the home range is its 'home', its main sleeping quarters or, in the female, the place where the young will be born. This home is psychologically the most precious part of the home range, to be jealously guarded. It is convenient, therefore, to have surrounding the home an area that is strictly private, from which other members of the species shall be barred, except at the breeding season, when a mate may be allowed to enter. This is the territory.

Territory. This can be best understood from a diagram, but the diagram can be misleading. The idea of territory was first formulated for birds and everyone is now familiar with the concept of a bird's territory with its invisible boundary rigidly defined by the use of visual landmarks. For mammals the situation is less well defined since their territories are demarcated by scent-marks, depending not on the eyes but on the nose for indentification. We, using eyesight, can see a boundary to a property even if there is no fence, provided there are one or two shrubs, trees or posts as landmarks. Birds do much the same for their territories.

In birds, also, it is customary to speak of a

A group of lionesses moving across the African savannah at the foot of Mount Kilimanjaro evidently starting out on a hunting foray. Lions live in groups called 'prides', each varying from as few as three to four individuals to as many as 30.

territory with a nest at the centre, and not to speak about a home range, although we now know such a thing is used.

In mammals the boundary of the territory is often ill-defined and that of the home range even more so. This is because a mammal uses trails to reach particular places in the home range. It marks these with scent so advertising its 'right' to the trails as well as the feeding points, bolt-holes and drinking places. The areas of land between the trails are neutral ground or may contain trails made by neighbouring members of the species. The result is that home ranges may overlap. Indeed, two or more of the same species may share one or more of the trails but they manage to do so at different times, so avoiding conflict. The domestic cat does this.

Although we use sight as the main sense it is easier to understand the wild mammals' use of living space by reference to human activities. In its simplest form, at least in western society, a man living in a house in a village has a garden around his house which is his private property. He may not fight every intruder but he looks questioningly at any stranger entering the garden. So we can call his house his lair and the garden his territory, and here he spends most of his time. For complete living he must go beyond these, to his work, to the shops, and so on, using an area that is used by all others in the village – the home range – but using it only to reach certain places and using specific trails to do so.

There is another feature recognized in the mammals living-space, known as the core area. This contains the best feeding and resting places and the best bolt-holes. While home ranges of two or more individuals may overlap, the core areas are usually sacrosanct or nearly so to one or a pair of individuals.

Knowledge of mammalian territories and movements within them, as well as methods of marking them, except where these are strikingly obvious, has hitherto depended almost on hearsay evidence, on accounts given by trappers and hunters especially. Some of this information has proved of value. Much of it has failed to be substantiated by more careful research during the last twenty years. In few instances has a complete picture emerged, with the result that the whole subject is slightly confused. The position is not improved by whether the species is solitary or gregarious, whether male and female live together other than at the breeding season, and a variety of other factors. We can only take a few of the better studied examples to see how the pattern varies.

Life Within the Territory. Shrews are among the most primitive of mammals. They are also solitary, males and females coming together briefly to mate, then going their separate ways. We should expect therefore that their territoriality should show a

Rabbits live in colonies, usually spoken of as warrens, but more properly named as buries. Within each bury the individuals occupy distinct territories shared by a few males and a larger number of females. The territory is jealously guarded against strangers, but on the whole its boundaries are respected, probably because they are well-marked by faecal pellets dropped at circumscribed sites and by urine. The secretion from chin glands in the males are rubbed on landmarks such as the bases of trees and posts.

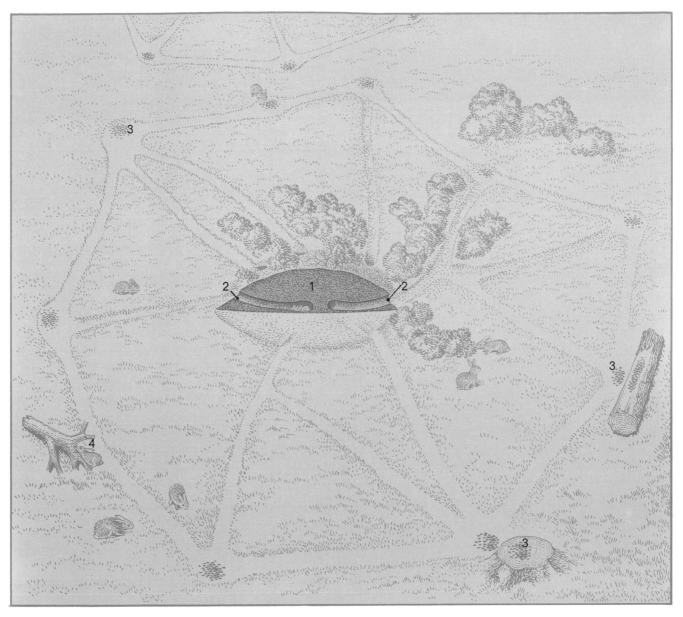

The territory of a colony of rabbits extends beyond the area of the warren or bury (1) and is characterized by a number of paths that lead to the bolt holes (2). The perimeter of the territory is marked by faecal pellets (3) or scent which warns off rabbits from other colonies. Scent is deposited on branches and other objects (4). This scent is secreted by glands under the chin of the rabbit, and is rubbed off during a process called chinning (inset).

primitive pattern, which it does. Each shrew builds a nest in which it sleeps for about half the day, not continuously but in alternating periods of about three hours activity and three hours rest, throughout each 24 hours. Around the nest is a territory which the shrews defend vigorously against others of its

85

kind. Beyond this is the home range in which it forages. The home ranges of several shrews may overlap and the shrews using these may use the same runs or tunnels. Should two of them meet, and this it would seem from circumstantial evidence may be relatively frequent, time and energy are not frittered in physical combat. The shrews indulge in vocal contests, squeaking at each other, until one retires.

At the other end of the scale for size, from the shrew 3 in (76 mm), is the lion 7 ft (2 m), an animal with a similar pattern of territoriality, although having a different social structure. Lions live typically in prides, of up to 35 individuals. The size and structure of the pride both vary. The size tends to be larger on the savannah, smaller in forested areas. It seems to depend on the supply of game. The structure varies for less predictable reasons. A pride seems to originate as a family group, with one adult male, several females and their young.

Lions make no nest and have no lair in the strict sense. They have a resting place and around it is a territory from which intruders are driven. Beyond this is the home range. Lions are indolent by nature and assuming game is sufficiently plentiful the ranges of two or more prides may overlap without the prides coming into conflict.

These two examples may be called the typical of the pattern for mammals. Those that follow represent some of the variations that occur.

A male House mouse, in size equivalent to the shrew, will occupy a nest with several females and will defend a territory around it vigorously. The home range is small, about 50 sq ft (4·6 sq m), and occupants of several nests will use if for foraging. The Mexican ground squirrel, a medium-sized rodent, whose habits have been closely studied is colonial but not social. Each colony comprises several burrows within a small area and the immediate ground around each burrow entrance is a defended territory. Each has its home range of 50–100 yd (46–92 m) in diameter, but other squirrels of the species are tolerated inside the home range provided they do not approach a territory.

The Uinta ground squirrel has also been extensively studied. Although closely related to the Mexican ground squirrel there are significant differences in its territorial pattern. It also is not social but it is gregarious like the Mexican species but with a difference. Uinta ground squirrels live together at close quarters but each is solitary, driving away others that show signs of coming too close, preserving individual distances from each other. The only time that close contact is tolerated is when male and female come together for mating, and this period lasts for a week only.

Wild Guinea pigs, another rodent, are gregarious and all occupy the home range equally except that there is a social hierarchy – commonly called a 'peck order' – and the dominants take precedence over the subordinates for the best food, and among males the dominants take the pick of the females. There is a similar situation for the Common or Brown rat. It is gregarious. Common rats live in large communal burrows and forage over a large home range. Among Common rats, also, there are dominants that take the best food and the pick of the females.

Territory in Rabbits. No single species of mammal, apart from the common rat, has been so extensively observed in the wild as has the European rabbit. Yet it has been necessary for several precise studies to be made on this common animal to obtain even a slight insight into the working of its territorial pattern. These studies have been made by several researchers in Britain and even more in Australia. They have been made on populations of rabbits enclosed within wire-netting fences, in areas of up to two acres. This is the first disadvantage, that the rabbits were then living to some extent under unnatural conditions. The second disadvantage was that the observers had to watch the movements of individual rabbits mainly through telescopes. These give a very restricted field of vision – a severe handicap when trying to keep track of many individuals moving about, at the same time trying to identify them as individuals, as well as trying to interpret what each is doing.

It is not surprising therefore that the results obtained by the various workers do not always tally as well as could be hoped. To begin with they do not all differentiate between a territory and a home range, yet the existence of the latter is implicit in their references to the distances rabbits will travel between the burrow and a feeding ground. It is implicit also in the references to the use of bolt-holes that are not their permanent burrows.

In attempting a succinct synthesis of the various observations set forth it seems that the older does

Mammals that live in open country, such as Dall's sheep ▷ pictured here, tend to have larger territories than do those of more restricted regions. Another important factor determining territory size is the abundance of food.

take the initiative in digging burrows at the beginning of the breeding season. Their activities are joined then by the bucks and younger does, and each doe then establishes a territory around the entrance to her burrow, which is jealously guarded.

There is, however, a considerable variation in the aggressiveness of the does, some being very quarrelsome, others very placid, some being so peaceable that there is no aggression at all between them. So they sort themselves naturally into groups of two to six, and in the breeding season one to three males attach themselves to each group, the number of males in each depending on the number of females present. There is an order of dominance and subordination among the males in a group, and to some extent also among the females.

Even if the details summarized here for the European rabbit are not always correct the broad picture presented is consistent with that found in other species. In the cottontail and the Swamp rabbit, both of North America, the pattern is basically similar although they are not burrowers, the females making a nest above ground. In all three it is the females that set the pattern for the organisation of the group and determine the site it occupies. The females also determine, by the degree of tolerance they show to other females, the numbers in the group.

Territories and Social Structure. Hippopotamuses differ enormously from rabbits in size and habits and in being amphibious. All the same there is a striking similarity in the pattern of territories. As we have already seen, in a river or lake inhabited by a colony of these huge pachyderms the females and young live in a central area of water, bounded on one side by the bank of the river or lake, and usually with a sandbank. The males hold separate territories in the water around the periphery of the females' central territory. A female usually makes a nest, of trampled reeds or other vegetation, at the water's edge, for the birth of her single calf. So the nest is transitory.

The territory is very definite, and for one male to enter the territory of another is to risk a terrific battle. Aggressive intent is signalled by the wide open mouth. In combat the two rise high from the water, with mouths agape and a tremendous flurry of water, each seeking to slash the other with its enormous tusks. Severe wounds are inflicted, which soon heal, but if one sustains a broken foreleg the injury is fatal. With such a huge bulk to support a hippopotamus with only three functional legs is doomed to die.

A male may enter the females' territory but on strict committee rules. He may only do so if all the females are lying down. Should one of them rise to her feet he must lie down or be set upon by two or three females. Although he has the superiority of

The typical pattern in a colony of hippopotamuses is for the females and young to live together in a central area along the bank of a river or lake. Around the periphery, each adult male has his area from which he jealously drives any other males.

The rhino like this Great Indian rhinoceros requires a large territory which the males contend with ritual fighting. Often, however, the rhino is nomadic and, in this case, fighting over territorial claims rarely occurs.

weight and strength over one female, he is no match for two or more. So long as he obeys the rules he may stay.

Since the social organisation of hippopotamuses is based on fighting, going ashore to feed, as they must do, would present difficulties if the feeding areas were not rigidly parcelled out. So to feed, each goes on land at the same point each time and follows a path which fans out onto the grasslands. All these feeding areas together form the home range of the colony, but each member of the colony has his or her well-defined part of it, and there is no overlap.

The eyesight of a hippopotamus is relatively poor. Boundaries and trails in the separate feeding areas are well-marked, as is usual in mammals, by an odoriferous substance. In the hippopotamus this is provided by dung. As the animal defecates its short tail beset with stout bristles is whisked around, almost with the speed of an electric fan, spraying the dung onto bushes.

The European red fox is a solitary species except during the breeding season when it lives in pairs or family parties until the cubs are independent. Their territorial behaviour shows the same three essential components. The nest for the cubs is underground,

under cover of windfall or among boulders. An area around this is the territory, defended especially by the male. Beyond this male and female forage and have bolt-holes and resting places. This home range may be shared by other foxes all of which mark their trails with scent from their glands, the characteristic foxy smell. Jackals, other members of the dog family, have a similar arrangement, but they tend to meet more than foxes and be more tolerant towards each other, especially around the remains of a lion's kill.

A troop of hunting dogs of Africa was studied. It consisted of six males, two females and fifteen juveniles. There seemed to be complete equality between the sexes with no sign of dominant or subordinate individuals among either males or females. The burrows, used by the females for nursery purposes, were surrounded by a territory of a square kilometre, from which all carnivores, including lions, were excluded. Around this was a hunting range, the equivalent of the home range, of up to 77 sq mi (200 sq km), but ranges of other packs of hunting dogs overlapped this at various points. Should two packs meet there is barking but no fighting, and the smaller pack retires.

The movements of wolves are comparable to

those of hunting dogs. This is one of the few statements that can be made about wolves with any great degree of confidence. Another is that they travel greater distances in winter than in summer when they are tied to their dens by the presence of young still being nursed. Presumably there is a defended territory around the den and the home range, or hunting range, may be twenty miles in any direction, giving a round trip of 40 mi (67 km) a day to feed adults and their pups. The trotting speed of a wolf is 5 mi (8 km) per hour, so 40 mi (67 km) a day is no great hardship.

In winter, with the young able to keep up with the adults, and with no need to return to a nursery den, a pack will range farther afield. Much observation has been made of these movements, in recent years aeroplanes and radar have been used to track them. Using these methods it has been shown that wolves may cover 124 mi (200 km) in a day in a hunting area of up to 2,000 sq mi (5,180 sq km) or more, often following migrating herds of caribou or domestic reindeer.

In fact, so many data have been collected that it is difficult to sort them out to make a clear and simple picture. Also the behaviour of the wolves varies greatly according to the terrain, the abundance of the prey and the season. What has been given here can therefore only be an over-simplified picture.

In one sense, the use of the term home range can be misleading. It can only be taken as an area of land enclosing the runways of a pack or available for the use of the pack. These runways are marked by urine on any rock, stump, log or bush, the males with one hind leg lifted and crooked, or urinating on the ground with all four legs slightly spread or with one hind leg lifted under the body. The females either squat or lift a hind leg under the body. Body rubbing may also be used, the animal rubbing against a solid object starting at the neck and continuing along the flank, then turning round to repeat the process on the other side. It may also roll onto its back and will do so on anything having a strong odour, a familiar sight in the domestic dog, which is assumed to have been derived from the wolf.

Lone Wolves. There are also lone wolves which seem to lead a nomadic life completely, merely denning up to rest by day and then continuing on, foraging, the next night. There may be several reasons why some individuals of a species break away from the normal and become wanderers. They may be outcasts or misfits, subordinates that have

been bullied, non-breeders or, in a manner of speaking, delinquents that cannot conform. Nobody is very clear about the causes, but since parallels can so often be drawn between the behaviour of humans and that of animals there is no reason to suppose there may not be drop-outs from animal societies as there are from human society.

In one set of observations on small mammals there may be a clue to at least one of the causes. The movements of small mammals such as voles and mice, or even weasels, can be studied by live-trapping and marking individuals. By continual live-trapping in a defined area the movements of the occupants of a colony can be plotted. In the course of time the investigator almost gets to know the individuals by sight. He becomes very much aware, as a consequence, of strangers that pass through the area. These have been named transients, and are wanderers which, so far as can be seen, have no fixed abode. The existence of wanderers, 'lone wolves', has been recorded but little more notice has been taken of them than that.

It is tempting to equate these transients with members of a settled human society that are unhappy unless they are forever moving around. Some become explorers, others drift from job to job moving about their native country or drifting from one country to another. They are said to have 'itchy feet'. The outstanding example, possibly, of an animal with itchy feet, although there probably have been many others less well-documented, was Huberta, the hippopotamus. This animal left St. Lucia Bay in Natal, where she was born, as soon as she became subadult, and for nearly three years wandered in a more or less direct line to beyond King Williamstown, in South Africa, a distance of a thousand miles. Her journey was fully documented by the local newspapers of the towns she passed through.

There has always been a mystery about the seals on Antarctica whose remains have been found 20–30 miles from the coast and 2,000 ft (610 m) up the glaciers. Captain Scott recorded finding a carcase, in 1903, 50 mi (80 km) from the coast – a remarkable journey for a seal – at a height of 5,000 ft (1,524 m) above sea-level. Whether these marine animals were prompted to make their overland journeys for the same reason that the kitten in 1950 climbed the Matterhorn or the Colobus monkey a year later climbed Mt Kadam in Uganda, is a matter for speculation. One explanation could be that they are examples of transients or

90

Shrews are insect-eaters and because of their small size need a high food intake to offset the loss of body heat. Estimates of the amount of food they eat each day vary from the equivalent of three-quarters to the total of their bodyweight.

animals with itchy feet that happened to have been noticed.

The Homing Instinct. Nobody could claim that Huberta got lost. If that had been so the map of her wanderings would have shown more erratic journeyings, of the animal trying to find her way back. Added to this many mammals have given evidence of a remarkable ability to find their way home over a distance.

There has been an accumulation of anecdotes about domestic cats and dogs finding their way back home over surprising distances. A typical example is of a dog taken to a new home by train, finding its way back by road to its old home, a distance of 20 mi (32 km) or more. Some of these stories seem to stand up to critical scrutiny and the few controlled tests that have been carried out suggest there may be more than a germ of truth in them. The subjects of such stories may be the more

gifted of their race in homing, and homing over shorter distances must be commonplace, although the means whereby it is done is still unknown.

In 1931, Bastian Schmid conducted a series of experiments with a dog named Max. It was three times taken to a place seven kilometres from home where it had never been before. It was taken there by closed van in a large wicker basket covered with tarpaulin, so excluding any possibility of its being able to recognize the route. Great trouble was taken to have observers at the point where the dog was set down and all over the countryside, to watch unobtrusively and report in detail all the dog did.

When set down in the farmyard, from where he had to find his way home, the dog appeared timid and suspicious, behaving almost like a wild dog. He scanned the unfamiliar countryside for a while, then stood gazing for a few minutes in the direction of home, which was hidden from him by a ridge and a wood. Gradually his irresolution wore off. He continued to face in the direction of home. Then, after half an hour, he moved off in the direction of home.

Max did not travel in a dead straight line, nor did he follow precisely the same route on each of the three occasions. So it seems that celestial navigation, such as insects, birds, fishes and toads are now known to use, can be ruled out. His reliance on landmarks must have been minimal, otherwise he would have followed the same route on each occasion.

There should be little surprise that dogs can find their way home in this uncanny fashion. Wolves are known, when carrying food back to the den, after making an erratic hunting journey, to travel for miles in a straight line. Even if we allow a familiarity with the home range this is still remarkable. It is much more likely that wolves are using a homing 'instinct' of which our knowledge is at present negligible.

Apart from the few experiments carried out by Bastian Schmid, there have been a number of other experiments, the results of which, only make the subject more baffling. For example, White-footed mice, of North America, were found to wander daily over a home range of not more than 150 ft (46 m). Individuals live-trapped and taken farther

away found their way home from distances of 500–700 ft (152–213 m). Exceptional individuals homed from distances of one to two miles, the last distance representing 75 times the greatest distance such a mouse is normally prone to wander.

Similar results have been obtained with marked Wood mice in Britain, and there was one instance of a marked European mole placed down a mile from its point of capture being trapped again 12 hours later near the original point of capture. We are left to speculate whether the mole travelled over the surface or underground. It may have done a little of each. Since this mole can travel at a speed of 2·5 mi (4 km) per hour over the surface, such a journey, from the point of view of time is not remarkable. It is the method used that is puzzling.

The Water vole of Britain is strongly resident and normally has a home range of 200 yd (188 m) along a river bank. Tested individuals returned from up to a mile away from home. A hedgehog found eating eggs in a hen house was taken and placed down a quarter of a mile away. The next evening a hedgehog like it (presumably the same individual) was found in the same hen house. Red paint was dabbed on its spines and it was taken and set down half a mile away. It was back in the hen house the next evening. When taken a mile away it failed to return.

Moles are solitary by nature and it is only under certain conditions that they share a common run. One of these conditions is when they are migrating, as when they move from a summer territory to a winter territory. For the most part, however, each occupies a discreet territory marked at the surface by well-defined groups of molehills.

The Timber wolf provides an almost text-book example of the typical pattern of territory-holding in mammals. Whether a lone wolf, family party or pack, there is a den with a small surrounding territory, which is aggressively guarded. No hunting is done within the territory. This is reserved for the home range, a much larger area than the territory proper. The home range is patrolled in search of prey, the wolf or wolves returning to the den to rest.

Experiments with bats have shown that many species can return to their roosts from distances of 20 mi (32 km) in a few hours or from distances of up to 200 mi (320 km) in a few days. The Spear-nosed bat of Trinidad was tested by being blindfolded. Individuals homed successfully from distances of up to eight miles. This suggests that bats use visual landmarks to some extent, and the times taken suggest random searching before getting on to the correct route. Even so, the performances of the Spear-nosed bat indicate that this is not the whole story.

Perhaps the most remarkable homing records for mammals are those for the Brown hare of Europe and Africa. There are several records in recent years of marked hares homing from up to 280 mi (450 km) from their point of capture. Reindeer are credited with finding their way home unerringly over distances of miles in blinding snow storms. This merely makes the problem more puzzling.

There is, however, an anecdotal record – that is, one not scientifically tested – of an Australian farmer who was out with his aboriginal helper looking for fresh pasture. Fifty miles or so from home they camped for the night and the Aborigine failed to hobble the horses properly. The next morning they were gone and the two men had some difficulty in reaching home. Months later he received two letters from stations over 600 mi (966 km) apart. Each said that a horse bearing the station's brand turned up. Both horses had gone back to where they were born.

There are people even in advanced societies and living in towns who find they can 'home'. Some of these when challenged have allowed themselves to be blindfolded, taken by car to unfamiliar country-side and set down, to find their way back unaided. The Pygmies of the Ituri Forest in West Africa and Siberian tribesmen have been tested and found capable of 'homing' to an even more marked degree, and to do so as a matter of course. When asked how they do it none of these people has been able to say. They merely ask, incredulously, 'Can't you do it?', or words to that effect.

What hope, therefore, have we of understanding this behaviour in animals if human beings possessing it cannot describe it? Yet, if there is a 'sense of direction' there must be sense organs or sense receptors in which it is lodged. These may yet be discovered. It is only in the past few years that the full equipment of touch receptors in moles has been appreciated. It has yet to be ascertained whether it is by the use of these, perhaps in combination with other sense organs, that give moles, and other full-time burrowers such as Mole rats, their extraordinary 'sense of orientation'.

Mole-trappers are accustomed to the way a mole will by-pass an obstruction in its run. It will drive a fresh tunnel around the obstruction and connect up with the tunnel the other side of it with absolute precision. A mole live-trapped in its run and taken 12 yd (11 m) away from it, and there released onto the surface of the ground, will immediately burrow out of sight. If suitable preparations are made it can then be trapped again in its original run a short while later. When the new tunnel is opened up and traced it will be found to run more or less in a straight line to the old tunnel. What is more, and this is quite remarkable, it will be found that the new tunnel has joined the old tunnel with precision.

It has been said that when the two excavating teams boring the Simplon Tunnel met, they were only a fraction of an inch out. This was a tribute to the skill of engineers and mathematicians using all the then known instruments and implements. Relatively, a mole does just as well, using a primitive brain to process data from sense organs the nature and existence of which we are ignorant.

This sense of orientation has been subject to exhaustive tests in the Mole rats of South Africa. The investigators used females with nests, which implies a greater urge to return home. As fast as the tunnel leading to the nest was blocked the female drove a secondary tunnel around the obstruction. She continued to do this, to reach her nest, as often as the investigators blocked the tunnel leading to it. And always the secondary tunnels connected precisely with the old tunnel. It was as though the Mole rats, instead of being blind, could see better than the investigators.

From Hermit to Herd

The use of territories is essential if species are to survive, and we can be sure not only that it is present in some form or other in virtually all species of animals, but that this has been so for most of their evolutionary history. It results in individuals of a species being spaced out to take the greatest advantage of the available food. The pattern for each species is also linked with the food supply. This is strikingly demonstrated by the way individuals of a species that inhabits both desert and savannah areas, are few in number and well spaced out in desert and live in large compact herds in rich pasture. Indeed, except in humans, it is a law that the size of any groupings of a species varies directly with the poorness or richness of the food supply.

In the previous chapter an examination was made of the territorial patterns of the smaller groupings, from the solitary species, living so much on their own except at breeding time as to justify the title of hermits, to packs or prides of no more than three dozen individuals in one group. Even in the largest of these the basis is the family, of monogamous or polygamous parents. There are, however, species of mammals that exist, or used to exist before man interfered with them, in enormous herds that covered the plains as far as the eye could see. They were mainly the large herbivores.

Dramatic word pictures have been drawn of the North American buffalo or bison that could be counted in millions. There are stories from the diaries and books of the first Europeans to visit subsaharan Africa of the countless herds of game. Remnants of these, large herds of zebra and antelope, are still to be seen in the Serengeti, but if the stories of the past are to be believed even these represent no more than a shadow of the vast herds that used to live, for example on the veldt of South Africa.

F. Martin Duncan, in his book *Wonders of*

Impala live in large mixed herds of hundreds during the dry season. During the wetter months the herds disintegrate into units consisting of one male and 15 to 25 females. The surplus males live in bachelor herds of 50-60. Sometimes the one-male units may join together temporarily, splitting up again later. They may at times associate with bachelor herds.

Migration, records the estimate of numbers of springbok made by a Dr Gibbons in South Africa. 'In the morning as soon as it was daylight we were out, and there we were sure enough, in a veritable sea of antelopes'. Dr Gibbons, looking at a nearby sheepfold, and being told it contained 1,500 sheep, estimated there were 10,000 springboks to an acre and that the huge mass of these animals that stretched for miles around on all sides must number 100,000,000 springboks.

This may or may not have been an exaggeration. Few people are alive today who could prove or disprove it. It is only possible to say that the numbers of springboks seen on that morning in 1888 must have been phenomenal to have evoked such a reaction. It is true they were on migration and therefore abnormally massed together. Yet even when dispersed for feeding they must have been very thick on the ground.

To examine the social structure of swarms of mammals of this kind we should start with another species that used to exist in these large numbers. We can then follow the studies of several observers of antelopes related to the springbok. The species we can start with is the Prairie dog of North America.

Prairie dogs and the Uganda kob. Prairie dogs used to inhabit the western plains of the United States in large numbers, in company with the vast herds of buffaloes. They are burrowing rodents about a foot long and their burrows cratered the prairies in vast colonies known as 'towns'. One Prairie dog town was estimated in 1901 to cover an area 100 by 240 mi (160 by 386 km) and to contain 400,000,000 Prairie dogs.

From 1948 to 1952 Dr John A. King studied Prairie dogs, marking their fur with patches of dye so that he could recognize individuals. He found that the burrows are not evenly spaced but grouped, and that a town is divided naturally into what he named wards and coteries. Each coterie is made up of a family consisting typically of a dominant male, a subordinate male, females and young, averaging about 18 in number with a maximum of 35 individuals. A coterie covers less than an acre, and several coteries make up a ward.

Dr King found that communication between Prairie dogs was by touch, smell, sight and voice. The first is important in sexual behaviour, play, fighting and mutual recognition, a coterie being a very closely-knit unit. The other three play a greater or lesser part according to circumstances, but touch is perhaps of immediate importance

between individuals. The most frequent demonstration of it is the 'identification kiss', which is exchanged whenever two individuals meet. Each turns its head and opens its mouth to permit contact with the other. The kiss may be given hastily – a sort of 'peck' such as humans use when they know each other well – as they pass each other when foraging or when they meet as they run together to a burrow. At times it may be prolonged and may end in mutual grooming, or one may roll onto its back still maintaining contact with the mouth, or the two may lie side-by-side, later going off to forage with their two bodies pressed closely together.

Should two Prairie dogs meet near the boundary of the coterie, or under any circumstances when their identities might be in doubt, they crouch on their bellies, wag their tails and crawl slowly towards each other until they can exchange the identification kiss. This is to make sure they both belong to the coterie. An individual straying in from a neighbouring coterie is driven off.

When a tame Prairie dog grooms the human hand it gives a tingling sensation to the skin, which is pleasurable and doubtless maintains the bond between members of the coterie. It may also serve as a skin massage and help in maintaining health. Male and female, parent and young, all indulge in mutual grooming. For the rest, Prairie dogs co-operate in nest building and in all other activities, including using the voice to warn of danger, whence the old idea that they posted sentries.

The social structure within a large herd is of a number of small groups in each of which females and young, together with several dominant males and a varying number of subordinate or bachelor males, occupy a home range for feeding purposes. This pattern was first studied in detail in the Uganda kob, a species of antelope, about a decade ago. What to a casual observer looks like a large herd stretched across the savannah is seen, on closer study, to be made up of a number of these units.

Within each home range is a central breeding area made up of 30–40 territories each occupied by a dominant male. These territories are small, 5–10 yd (4·6–9·2 m) across, and there may be larger territories near the periphery of the home range, five to

When two Prairie dogs meet they exchange a recognition ▷ 'kiss'. When one of the two is a juvenile this is the prelude to the adult grooming it, nibbling its fur all over. Such community living has been taken to extremes by prairie dogs.

six times as large as those at the centre. The significance of these larger territories is not clear since they may be absent and anyway mating takes place mainly in the central territory. It also appears that the pattern of territories varies according to the density of the population in the home range.

In any event, it is those territories at the centre that provide the focus for breeding and it is what takes place in these that is so highly interesting. Indeed, it is not putting it too strongly to say that the film taken of the Uganda kob's breeding marked a turning-point in the thinking of many zoologists. In this film a male kob can be seen restlessly patrolling the perimeter of his small territory. He may leave it from time to time to feed or go to drink, as do other males in neighbouring territories. There is no conflict between the males unless one of them inadvertently, in going to pasture or to the waterhole, oversteps onto the territory of another. Then there is a brief display of aggression between them, ending in the intruder going away. The only real fighting, and that is more a trial of strength than a bloody encounter, comes when a young male reaching full sexual maturity tries to oust an occupying male from the territory he occupies.

The male in possession, as has been said, patrols the territory restlessly, obviously in an excited state and clearly at the peak of breeding condition, as shown by the occasional emission of semen while no females are near to stimulate it. Then a female enters the territory and drops to a resting position at the centre, with her neck held vertically. She is placid to all appearances and she seems to take no notice of the male, nor he of her, as he continues his nervous pacing around the territory. After a while another female may join her, dropping down beside her and lying close to her.

There is nothing in the film, or in the written accounts of those that have watched the breeding of the Uganda kob, to reflect the old and firmly-established idea of two males fighting for possession of the female. On the contrary, the onlooker gets the clear impression that the male's whole concern, and all his attention, is concentrated on maintaining his territory intact. The females, for their part, are entering the territory of their own volition, making their own selection of the male that shall eventually sire their offspring. In other words, it is the female that does the choosing.

Another interesting point is that once a female has entered the territory other females are likely

The normal habitat of impala is the woodland edge, an area of transitional vegetation from which they can readily reach grassland to graze or thicker woodland for refuge. Ideally there should be water nearby.

Male Jackson's hartebeest sparring, a common habit which outside the breeding season is more a trial of strength than actual combat. The dominance structure of the herd is well-marked and determined by these trials of strength. The boss male marks his territory by what is called static-optical demarcation, that is, he stands on a high mound, such as a termite nest, and rules the territory that can be seen.

to choose that one rather than a territory that is without a female.

When one turns from contemplation of these observations and reviews all that is known about animal courtship, the real truth seems to dawn. Males do not fight over females, or if they do it must be exceptional. They fight over a territory, or the equivalent of a territory, making it easier for the female to decide where and with whom she will mate. She may, as a natural consequence, choose the most virile male, and therefore the most desirable from the point of view of posterity. It is the fact of males fighting for territory with a female standing nearby that has deluded observers in the past into thinking that the males are fighting for the possession of the female.

The Gazelle and Waterbuck. Researches into the social structure of other antelopes shows that the herds are made up of similar groupings to those found in the Uganda kob. One of these is Grant's gazelle.

This is a common medium-sized antelope of eastern Africa. It lives in groups each occupying an area about 300 yd (274 m) in diameter. Within this area are twenty or so individuals in three groupings. At the centre – the centre being psychological rather than physical – is the male with several females with their juvenile young, occupying a territory defended by the male. At the periphery of the home range may be a group of bachelor males. Near the dominant male and his harem are grouped females with infant young. When a female is about to give birth she goes apart from the others. Then, when her baby is a few days old she joins other females with infant young. It is only later, when the infant is becoming juvenile or subadult, that she re-enters the central territory of the dominant male.

The related Thomson's gazelle, even more common than Grant's gazelle, has a similar social structure. The main difference being that females with young tend to stay apart from the central group for a longer period.

In the waterbuck there is another variation in the plan of defended territory and home range. In this species the breeding territories are sharply demarcated and are located along the banks of a river, each territory having a length of the river bank as one of its boundaries. The home range extends outwards from the river bank and tends to be ill-defined and the boundaries of the breeding territories are also less well-defined and less vigorously defended away from the river. So instead of central breeding territories surrounded by the feeding grounds representing the home range, there is a linear distribution of breeding territories along the river bank that are readily recognizable, while the home range is less easy to make out. This is related to the waterbuck's preference for marshy places and for the long grass growing there for feeding. It is as if the farther the ground is from the river, the more the waterbuck lose interest in it, so the dominant males hug the banks, the females feed in the grass beyond it and the bachelor males must take pot-luck well away from the bank.

In writing about social organizations in animals, there must of necessity be a tendency to idealize the situation in order to produce a diagram that will be comprehensible and to avoid a lengthy description that will confuse the reader. In the case of the

waterbuck, for example, the actual disposition of the breeding territories, the home range and the siting of the bachelors will depend very much on the numbers within the herd and the state of the feeding grounds, the second of these being the most important. Where lush grass grows from the edge of a river or lake right up to the edge of a forest, as sometimes happens in East Africa, the breeding territories are not consistently sited along the bank.

In contrast to the defined breeding territories held by the antelopes so far discussed, the impala form breeding groups with no fixed territory. The dominant males attract females around them and they also herd them if they show signs of scattering. They do not define specific areas, the breeding territories being wherever the females are gathered together. There is, however, the same driving away of adolescent males to form bachelor groups and bachelors trying to interfere with the female groups are driven away by the male in possession. In this species also the sizes of the various groupings varies considerably. There may be two females with one male or 200, and a bachelor group may number anything from two to 60.

Shifting Territories. The social pattern also

◁ At the onset of the rut, in the autumn, the male herds break up, individual stags spacing themselves out and roaring to warn off other stags. This giving voice continues but with diminished intensity as the hinds are rounded up, each mature stag gathering as many as he can into a harem. He patrols the periphery incessantly to keep other stags, particularly the young stags reaching maturity, from entering the harem.

Male Defassa waterbuck *Kobus defassa* in Uganda. ▷ It is like the Common waterbuck in size and habits and is distinguished by the white round the eyes. The dominant males each occupy a territory of $\frac{1}{4}$ to 1 sq mi (0·7–2·7 sq km) along the riverside. The younger males occupy less well defined territories along the periphery of dominant males' territories. The females gather in herds of up to 30 on feeding ranges approximating to the territories of the dominant males. They are, however, not so tied to a territory as the males.

becomes less well defined in species that wander from one pasture to another. One of these that has been studied is the Brindled gnu, or Blue wildebeest. This is a large ungainly-looking antelope that ranges from South Africa northwards to Kenya. It is especially abundant on the open plains in the northern parts of its range. Within this species two types of organization have been observed. In one, the herd is sedentary, or resident, the plan is then much as in the waterbuck, Uganda kob, and others already discussed. The herd occupies a feeding area, the home range, and within this there are groups of females and young, males holding breeding territories and bachelor groups. In the second, the herd is constantly on the move because it is living in an area where seasonal changes in the vegetation are strongly marked. As the herd moves from one feeding ground to another it splits into temporary breeding groups made up of a male and several

females and young. Should the group be large, of 150 individuals or more, it may contain two or three males which not only tolerate one another but may assist one another in defence of the group.

In this second type of organization, there is a shifting breeding territory instead of a stationary one. Otherwise there is no difference in organization as between the nomadic and the sedentary herds.

Among the horses, including the wild horse, zebra and asses, there are three types of organization. In asses the sexes remain separate except during the breeding season. The females and juveniles are grouped under the leadership of an old female, the males being solitary or form at most small groups. For breeding, the males enter these groups and each mates with several females, the number depending on the fighting qualities of the male, not in competition with other males but because of the belligerence of the females. Such courtship as there

A herd of gnu, or wildebeeste, on the African grassland. Outside the breeding season, when the mature bulls form harems, the gnu are scattered over the plains during the wet season. During the dry season they move through the bush, along the rivers, sometimes in huge numbers, seeking out the new grass produced by local showers.

A typical scene on the East African savannah, a mixed herd of wildebeeste and zebra. In such a herd the two constituent species share the advantages of their different ways of sensing danger. The hoofed animals take their deep sleep in turns, so that some members of the herd are always awake.

is consists of fighting between male and female.

In zebras, the herds are made up of family groups of one male and one to six females and their young. These family groups are led by a female, the male following in the rear when the group is on the move, or walking to one side. Occasionally he may walk ahead. The young males leave the group in their second year to form bachelor groups and later they collect their mates from existing groups or when an adult male is growing feeble. The groups are stable to the extent that, for the most part, the females keep together with the same male until he is old and feeble. Even where there are large aggregations of zebras forming an apparently uniform herd across the plain it is, in fact, made up of family groups, each retaining its separate identity, with the bachelor groups also remaining distinct.

The only wild horse now extant is Przewalski's horse of Central Asia. This lives in large herds with a single stallion and many mares and foals. As the young males grow up they are driven to the fringes of the herd until they can either abduct some of the females to start another herd or fall victim to a predator. Observation of herds of domestic horses shows the same pattern. The dominant stallion has a heavy task since he defends from predators and watches over the herd generally as well as serving many mares. It seems only just, therefore, that he should have the first pick in feeding and drinking, a characteristic more familiar in the lion.

The organization found in asses is not very different from that of elephants. In both the sexes are separate except during the breeding season. The female elephants and young occupy one home range, the males living in another. When the females are in oestrus the two coalesce to form one large herd. In this, the mature males mate with the mature females and on the fringes of the herd the bachelors fight, presumably to establish some form of hierarchy. The fighting is bloodless and not sustained, and it seems more a test of strength. It may even be, although this must be regarded as no more than an expression of opinion, due to a growing virility with a need to 'test the muscles'. The result may be secondarily to establish a hierarchy. Certainly, the practical and observable result is that the victors in such bouts are as likely as not to enter the herd in this sexual jamboree, which lasts no more than two to three days, to seek mates among the younger females. When the love-making and fighting dies down, the males move away in their separate herd and the females and young are left to continue their normal feeding.

The more typical organization is seen in cattle and antelopes, in which there is a home range with central breeding territories. One modification of this is seen in the vicuña, the wild camel of South America. Its home range is also the breeding territory in which are about eight females and their young, with a single dominant male, the bachelor males being relegated to the fringes of the territory. The bachelors have to make do with the poorer grazing so ensuring the best pasture for the females and young – and the hard-worked master male.

Deer and Territorial Defence. The Red deer is a herd-forming member of another family of hoofed herbivores. Its behaviour can be readily studied because herds are so often kept in parks. The true habitat of this species is woodland, and it is found in temperate Eurasia and North America, through-out which it more often goes under the name of wapiti. In other species, for example in some of the

103

antelopes, it is known that the sizes of herds, and to some extent their behaviour, differs according to whether they are living in forests or on the open plains. In Scotland, denuded of much of its forests during the Industrial Revolution, Red deer live more usually on moorland. That is where F. Fraser Darling made his early and classic study of their behaviour. There, and in parks, the females and young herd together in a home range and around such a home range there will be groups of males. Towards the end of September the tranquility of the female herd is broken, at first by the roaring of the males, who later partition the females.

In a typical instance a stag manages to segregate a group of females and keeps them together by racing around them, tearing the ground with his hoofs and so creating a circle obvious to the onlooker's eyes. While thus patrolling he is in a state of excitement, roaring from time to time, keeping an eye on any female showing a tendency to wander, and also alert to any challenging male that may approach.

Even although the perimeter of the breeding territory of the Red deer may be obvious to the eye it is not necessarily a stable territory. The size of the harem within it may change as females are abducted or more females join it as other males become spent.

It looks on the face of it as though the stag is defending the harem rather than a territory. If this is so he is singularly unsuccessful. While he expends enormous energy in racing round the perimeter and expending time in driving off one challenger after another, other younger stags sneak in from time to time and mate with one of the hinds. This is especially true if the harem is a large one. The impression then is that the stag is more concerned with defending his territory than with the business of mating.

The Roe deer's behaviour is at the other extreme. The social organization is one of family parties. Most of the year roe are seen singly or in pairs, with family parties forming as the fawns grow old enough to accompany the parents.

◁ Red deer hinds fighting. Although mixed herds may sometimes keep together throughout summer and winter, it is more usual for the hinds to form matriarchal herds outside the breeding season. These are, for the most part, harmonious but fighting does occasionally break out. This is not fighting over territory but a matter of settling the order in the hierarchy of the female herd.

A herd of Red deer in October. ▷ During autumn, each stag associates with a harem of females which, as this picture shows, are smaller than the male and lack antlers.

Social Organization in Primates. One other group of mammals in which large aggregations of individuals are found are the primates, the monkeys more especially, including the baboons. Most primates, including the apes, live in forests where they have both vertical and horizontal living space, as well as an abundance of food. As a result there is little competition between neighbouring troops and little aggression. Should foraging troops meet there is a fair amount of vocal expression and grimacing but seldom actual fighting. It is only in captivity, under restricted conditions, that real fighting occurs.

There are a few primates that inhabit open ground, including the Patas monkeys and the Hamadryas and Chacma baboons. The troops of Patas monkeys are generally not large, usually numbering about 15, rarely exceeding 30. Yet the familiar pattern is there, of a dominant male, a number of females and young and bachelor males on the outskirts of the troop.

The Hamadryas baboon sleeps at night on cliffs or rocky out-crops in large numbers. By day, the troop splits up into parties to go foraging. Each party includes an adult male, several females and their young, based on the polygamous family party. Bachelor parties attach themselves to the one-male parties and sooner or later each young male mates with one of the adult females. In due course one of the females goes off with a young male to form the nucleus of another group.

When out foraging, the females in a party keep close to the dominant male. They have to do this or take the consequences. Their tendency at first is to stray but the male teaches them not to by biting the neck of any female not conforming.

The Chacma baboon moves about in troops of 30 to 200 individuals. Each troop contains females and young with several adult males and young males on the fringes. The adult males establish a hierarchy best seen when a female is at the height of oestrous. The dominant male has priority of access to her.

In primates, therefore, the social organization shows the same range of variation seen in the large herbivores, and where the groups are large there is a similar structure in the group except that territorialism seems to be lacking. This may be something of a delusion fostered by the human observer's approach to the subject. It seems that only where a territorialism is apparent to the naked eye does the human observer recognize it. Yet the Hamadryas baboon keeps his troop of females within bounds by fear of chastisement as certainly as the Red deer

stag keeps his harem within a breeding territory by rampaging around it and cutting a visible circle with his hoofs.

In conflict with what was seen earlier in the Uganda kob, in which there seems definite evidence of choosing her mate, in the brief descriptions of the behaviour in other species there tends to be the impression that more often the females are totally subservient to the will of the male. It remains a question, therefore, whether the Uganda kob female is exceptional in exercising choice.

Animals with Sex-Appeal? The bushbuck, a medium-sized antelope of Africa, lives in forested regions and, as is usually the case, does not form herds. It lives in groups of a female with young and a male. Bushbuck feed in clearings which constitute their home range, their resting territories being in the undergrowth and trees surrounding the clearing. In the evening they come out to feed. R. Verheyen has described the events that take place in one such clearing and although he did not witness a mating he described how, whenever a particularly handsome male appeared he would be joined by one of the females, who would then accompany him back to his territory. Love is a subtle affair and this one incident is slender excuse for speculating on how far females in the herd species are capable of more than merely being rounded up by a dominant male.

It is some years now since somebody studying the courtship of Fiddler crabs noted that there was one particular female that seemed to catch the attention of all the males the moment she came out from her burrow. About the same time, and quite independently on the other side of the world, another investigator studying courtship in cuttlefishes also recorded a female who, wherever she swam by, seemed to act as a magnet to all the males, even those that were already courting other females. In both these scientific reports there seemed to be the implication that occasional individuals, even among animals, have the elusive quality of sex-appeal.

There is the story of a farmer who owned a pedigree cow that spent much of her time nuzzling a bull over a fence. The cow was destined to be served by a pedigree bull and was taken for this

A herd of topi in East Africa. Topi are a subspecies of ▷
Damaliscus korrigum, one of the Bastard hartebeestes.
The Bastard hartebeestes are smaller and less ungainly than
hartebeestes, often with rich plum-coloured, short-haired
coats. All have shorter faces than true hartebeestes. Their
horns are prominently ringed, spring separately from the
skull and curve evenly backwards instead of rising in
angular shapes from a common pedicle.

A troop of Olive baboons spread out while foraging, ready to move to cover if danger threatens. Each troop is a tightly knit unit and individuals seldom change troops. Dominant males and females with young travel at the centre of the troop when foraging over the home range. At the end of the day the troop retires for the night to a core area consisting of a clump of trees.

purpose three times to that handsome beast. Three times she conceived. Three times she dropped her stillborn pedigree calf. In desperation, the farmer, seeing her once again nuzzling her chosen, low-born bull over the fence, opened the communicating gate and let her into the field with the bull of her choice. She had several calves after that, none stillborn.

The cow-bull episode may or may not have been pure coincidence. Linked with the suspicion of sex-appeal in a Fiddler crab and a cuttlefish, it should act as a warning. The study of social organization and sexual habits in mammals has been going on for less than 20 years. The knowledge we have of both, a summary of which is attempted in these pages, is little more than a preliminary to laying the foundations of the full story. When, in the future, the mature edifice begins to rise from these foundations, it may present a very different picture.

Hibernation

One way of escaping temporarily the effects of stress, anxiety or the rigours of the climate is to go to sleep. Many animals do precisely this. In temperate climates they avoid the cold of winter by going into hibernation, sometimes called winter sleep, a misleading if convenient term because it is not true sleep. In hot climates extremes of heat are avoided by a similar process, known as aestivation.

Relatively few mammals use either of these, but because they are well-known processes, and because we are only now beginning to understand them, they are worth examining in such detail as we have.

Mammals, like birds, are homoeothermic, or warm-blooded, which means they can control their temperature to keep it at a more or less constant level, despite the fluctuations in the temperature of their surroundings. All other animals are cold-blooded, or poikilothermic, their body temperature fluctuating with that of their surroundings. There are a few mammals that can become temporarily poikilothermic, and mostly they do so when hibernating. Dormice, hedgehogs, marmots and some squirrels living in temperate latitudes regularly sleep the winter through. The most confirmed hibernants are, however, the insect-eating bats of these latitudes. A very few species of bats migrate south, but mostly bats find their way to caves or cave-like places to pass the winter.

It is now the custom to speak of poikilotherms, such as reptiles, entering a winter torpor, and insects are said to over-winter. Bears are always spoken of as hibernating but although the heart beat falls from 40 a minute to 10 or less, theirs is a true winter sleep and is now referred to as winter dormancy. The she-bear delivers her cubs while dormant. She rouses just sufficiently to bite the umbilical cord to free them after birth, and to lick them clean of birth membranes and fluids. The newly born cubs must suckle but because they are abnormally small they do not make heavy inroads into the food reserves of the mother. Newborn bears are no bigger than rats. Those of a Grizzly bear

It is usual to think of bears hibernating or at least denning up for the winter. This is true only of pregnant females. They den up to give birth and, even so, do not truly hibernate. Their winter sleep is now known as winter dormancy. Males remain active through the winter, like the one seen here watching a puma at its kill.

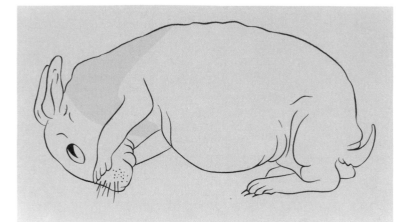

Ordinary white fat can be metabolized to produce heat. When brown fat was first discovered it was called hibernating gland, because it was most obvious in mammals that hibernate. Then it was seen in young mammals, as well as some adults, of non-hibernating species such as the young rabbit, seen here. It is now known to generate heat far more rapidly than white fat, and has been described as a sort of electric blanket. The full significance of brown fat has yet to be worked out.

weighing 700 lb (318 kg) weigh only 1·5 lb (0·68 kg).

The European badger is also said to hibernate in the more northerly parts of its range. This is, however, a carnivorean lethargy, as in bears.

Changes during Hibernation. The difference between winter torpor and true hibernation is that the former is due to a lowering of the body temperature due to a drop in the temperature of the surrounding air. In hibernation there is a series of internal changes which prepare the body for hibernation. Then the drop in temperature provides the stimulus to enter into hibernation. For example, one change is an increase in the amount of insulin, and an animal that normally hibernates can be artificially induced to do so out of season by injecting it with insulin and putting it in a refrigerator.

In the American ground squirrels an internal rhythm is known to determine the time of entering and awakening from hibernation. When kept experimentally in constant conditions of temperature and day-length, a Ground squirrel will go into hibernation at the same time as those still at liberty and subject to the onset of winter conditions.

Presumably there is a similar internal clock in others, although it has not been demonstrated. It is observable that bats are no longer seen flying on their usual beats in mid-September even when a late warm spell makes the air, even at night, as warm as in mid-August. On the other hand, towards the end of summer, hibernants feed voraciously, accumulating fat in the body, and this may be the determining factor. Bats, for example, become very fat then.

In the autumn the daily sleep period of the nocturnal hedgehog increases, the periods of activity grow progressively less. At the same time the body temperature during daily sleep gets lower and lower. All this suggests a running down of a cyclical rhythm.

Whatever the reason, at the appropriate time in autumn bats make their way to caves, often flying long distances to do so. Caves give the humid atmosphere necessary to them since they must conserve moisture during hibernation. There will be a steady loss by evaporation from the lungs, although breathing, in common with all bodily processes, is drastically reduced, unless the surrounding air is moist. Sometimes conditions in the caves are so moist that dew forms on the bats' fur.

The normal breathing rate for bats is about 200 a minute continuously. When hibernating this drops to 25–30 a minute for three minutes, then there is a pause of 3–8 minutes without breathing. In some species there is so little breathing it can hardly be detected.

Some bats have been shown to live to 27 years and it can be suspected that this is not the maximum longevity. This contrasts with the short lives of non-flying mammals of similar size. In shrews the maximum is 15 months. Harvest mice, in captivity, reached five years, by which time they not only showed signs of advanced senility but also behaviour more appropriate to their infancy: a 'second childhood', in fact.

The long life of bats must surely be linked with their sleeping habits. They hibernate for half the year and they sleep for nine-tenths of the other half. Most bats hunt for an hour or two at or about nightfall and for the same time just before dawn. During that time these use a high level of energy, offsetting this by falling into a torpid state during the day and between the two hunting sessions when they use only a twelfth of the food or stored fat that they use in active flight. If a colony of bats is disturbed during the day many of them will wake immediately, even fly away, but others will take 20 minutes or more warming up before they can fly.

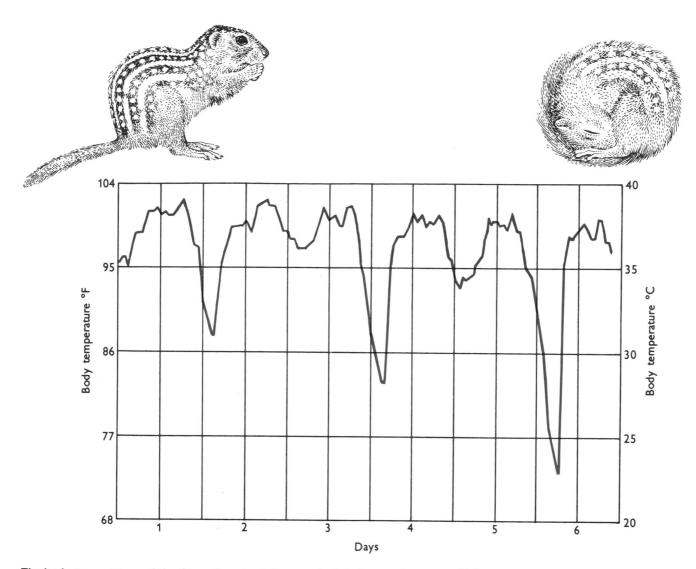

The body temperature of the Ground squirrel drops each night but, at the onset of hibernation, these become progressively lower until torpor sets in during the fifth night.

This long period for arousal shows that the daily sleep is akin to hibernation, and the fact that many are aroused straight away suggests that it is less constant than in hibernation. Nevertheless, even the hibernation is not unbroken. If a bat has chosen a part of a cave where the temperature drops at times too low, it will fly to another more suitable spot in the cave. Bats have also been recorded leaving a cave in mid-winter and flying to another 20 mi (32 km) away.

The Hazel dormouse also falls into a torpor resembling hibernation when it is not actively foraging. Indeed, the moment a dormouse stops moving, even when on a feeding expedition, it looks as though it has fallen asleep. Even so, its record for longevity is only five years.

The internal changes in preparation for hibernation concern mainly the blood. The organs in the pancreas known as the islets of Langerhans which produce insulin increase in size. This insulin induces other changes. The blood-sugar is reduced, ultimately to half the normal. Magnesium is released into the blood from the body cells and this, it is believed, acts on the temperature regulating centre in the hypothalamus, an area in the floor of the brain. Eventually the magnesium content of the blood becomes double the normal.

The hibernant seeks a sheltered place, sometimes a special winter nest, as these preparations begin. Although it will be cold when hibernating it will not survive temperatures below a certain level. So it must lie protected from the weather. The tolerable temperatures for most hibernants lie between 38–50°F (4–10°C). A very few species can tolerate temperatures below 32°F (0°C). For all, if the temperature drops below the critical level, the hibernant must wake or perish. Temperatures above the critical level tend to awaken it.

Other physiological changes associated with hibernation include a reduction in the activity of ductless glands, such as the thyroid, parts of the adrenals and the front part of the pituitary lying under the brain. These are glands that secrete hormones, or chemical messengers, into the body fluids, especially the blood. The anterior pituitary normally sends hormones to the thyroid as well as the adrenals, and these in turn secrete hormones that increase the metabolic rate, the rate at which the life processes work. The parathyroids, small glands near the thyroid, increase in size but the significance of this is not yet known. There is a marked migration of white corpuscles from the blood vessels to within the wall of the stomach and intestines, sometimes in such numbers as to form a gland-like mass. Nine out of 10 white corpuscles leave the blood to take up station somewhere else in the body. The white corpuscles are bacteriophages, that is they consume bacteria, and presumably their re-distribution gives a defence against bacteria that could develop from remains of food in the stomach and from food-waste in the intestine, remaining at the time hibernation commences, although evacuation of the bowel prior to hibernation gets rid of most of these. There is an accumulation of lymph tissues around the blood-vessels and the activity of the kidneys decreases, which reduces water loss from the body.

The winter sleeping nest, or hibernaculum, of small mammals other than bats, may be of grass or dead leaves, situated underground or in a sheltered cavity, among leaf-litter or in a hollow tree. In it the hibernant rolls up to cut down the loss of heat from the body. In the Hazel dormouse, for example, the chin rests on the chest, the fore-paws are clenched on either side of the chin, the hindlegs are brought forward so that the hind-paws touch the chin and the tail is wrapped over the face and head and onto the back. The eyes and mouth are shut and the ears folded down onto the sides of the head.

Although the body of a hibernant feels deathly cold to the touch the heart remains warm, the temperature of the body falling progressively from the heart to the surface. Presumably the brain must also remain near normal. Also, there is a special dilatation of the capillaries so that the blood pressure does not drop. The effect of this is that when the moment for arousal comes, at the end of hibernation, the animal can quickly warm up again. In both the cooling of the body and the warming up, the temperature control is in the hypothalamus.

Incomplete Hibernation. It used to be thought that hibernation was a period of continuous somnolence. In some species it is nearly so, but even in the dormouse there are signs that the animal wakes at times and feeds on the store of nuts and berries it had laid in. In the hedgehog the hibernation period lasts from mid October until late March or early April. Some individuals sleep throughout this period but others sleep intermittently until December or early January. It seems that sleep is less profound in the subadult hedgehogs, until the end of the year.

Common or Hazel dormouse hibernating, tightly curled into a ball, a means of preserving heat. Its winter sleep is so profound that, taken from the globular nest of grass among the leaf litter, it can be rolled across a table top without waking. This deep sleep may be protective, against the weather, but the fact remains that winter is the period of highest mortality in this species, from predators.

The dormouse seems most of its time to be asleep or about to fall asleep, although it can move with agility whenever the occasion demands. It is also one of the most complete hibernators. This is something of an enigma. The usual theory is that animals hibernate to tide over a period when food is short. Yet for the dormouse, the winter season is a period when nuts and berries, its main food, are most abundant.

Bats, as we have seen, may wake during hibernation and, in recent years, it has been shown that they feed during hibernation. Examination of the droppings on the floor of the cave, where Horseshoe bats are hibernating, has revealed the presence of fresh remains of the wing cases of Dung beetles. It was then found that the bats sometimes went out to feed on these but only when the air temperature reached 50°F (10°C).

Not only does the duration of hibernation vary from one species to another it can also vary from one individual to the other within a species. In the California ground squirrel many of the adult females hibernate from the beginning of summer until December or January, the males and the young having much shorter periods. This species has also been the subject of much research. Through it we know that arousal from hibernation throws a great strain on the whole system, and especially on the nervous system. On arousal, the heart which has been beating at 3–5 times a minute is stepped up to 400 times a minute, within 20 minutes of the animal waking. The intake of oxygen during hibernation is only 20 millilitres and this increases to 200 times this amount within the first half-hour. This means that the squirrel uses nine times the amount of energy in that half-hour than it uses during $5\frac{1}{2}$ months of hibernation.

The other experiments on this Ground squirrel consisted of injecting dogs, cats, rats and mice with a small amount of fluid from its brain. They fell into a sleep resembling hibernation, with a marked drop in the body temperature and a lowering of the breathing rate and the pulse. They were revived by warmth similar to that experienced by hibernating animals in spring. Natural hibernants seem to have a mechanism that switches on consciousness when the warm spring weather arrives. So far the mechanism causing a' hibernant to rouse is little understood. The rising temperature of the air around the hibernaculum has some influence, but more important is the brown fat lying around the region of the chest, shoulders and neck.

The Function of Brown Fat. Ordinary fat is white but in 1551 Konrad von Gesner, the Swiss naturalist, was dissecting the carcase of a marmot when he came across a mass of brown tissue between the shoulder blades. A similar brown tissue was later found in the bodies of other animals, especially in those that hibernated. It was therefore called the hibernating gland. Its function was, however, not clear and the puzzle was intensified when brown fat was found in rats that do not hibernate and also in newborn animals. Then in 1963, both at Oxford University in England and at Harvard Medical School in the United States a thorough investigation was made. It subsequently transpired that this was a kind of electric blanket.

When an animal baby is born it leaves the warmth of the mother's body for an atmosphere that is cooler. An adult animal suddenly exposed to cold can increase the warmth of its body by shivering. A newborn animal cannot shiver. The brown fat, in effect, switches on the heat.

The Golden-mantled ground squirrel *Citellus lateralis*, of western North America, lives at altitudes of 6,000-13,000 ft (1,830-3,960 m) above sea-level. Typically, Ground squirrels hibernate, but with them as with some other species, much depends on the latitude at which they live. The Golden-mantled ground squirrel, for example, does not hibernate in the southern part of its range.

Ordinary white fat helps to insulate the interior of the body, so keeping it warm. The cells of white fat contain only one droplet of fat each. The cells of the brown fat are larger, have more droplets of fats and many more mitochondria than those of white fat. Mitochondria are very small bodies, visible only under the high powers of the microscope. They burn fat to produce heat. The more there are, the more quickly is the fat burned, or oxidized, and brown fat produces 20 times the heat of white fat in a given time. Moreover, as the temperature of the air drops, so its production of heat is accelerated, as if it were controlled by a thermostat.

When the temperature of the surrounding air falls, the brown fat warms the hibernant just enough to prevent it from dying. If the temperature falls still lower, it warms the animal up so that it wakes, becomes active and can seek a more sheltered spot. Studies on an American bat have shown that when it is waking from its winter sleep its brown fat is 8°F (4°C) higher than the rest of the body and in a quarter of an hour it warms the whole bat to its summer normal.

What sets the brown fat to work is less easily told. We can only say that probably it is the hypothalamus, possibly working through the pituitary as a result of various external stimuli.

So far, only hibernants of the north temperate regions have been mentioned. There are others but information about them is scanty. The Australian duckbill or platypus has been said to hibernate but this is not certain. The echidna, or Spiny anteater, also of Australia has a body temperature that fluctuates between 84°F (29°C) and 90°F (32°C), which is lower than most mammals and more like the temperatures of reptiles. Both the platypus and anteater have a number of other reptilian features, especially in their skeletons. So long as they are feeding well, echidnas can tolerate temperatures down to 5°C. If they then stop feeding they hibernate, their body temperature falling to a little above that of the surrounding air. Dwarf lemurs of Madagascar are said to hibernate and Mouse lemurs also, although in their case it is more a resting state during which the animal feeds on the fat in its tail.

The most complete hibernants are the Ground squirrels of Point Barrow on the northern coast of Alaska. In the arctic region snow persists for most of the year and the ground is permanently frozen to a depth of several hundred feet. During the short summer the earth may thaw to a depth of a few inches, at most a few feet, especially in areas of good drainage where the sandy soil is raised in hillocks, forming 'islands'. On these the risk of flooding is low, and there the Ground squirrels have their burrows.

The Barrow ground squirrel sleeps for nine months in the year and has to pack all its activities, breeding, rearing the young, laying in a food store and accumulating fat in the body, in three months. The young are born towards the end of June. Their eyes open at the age of 20 days and two days later they start to dig their own burrows, find their own food and fatten themselves to go into hibernation by the end of August or beginning of September.

A striking feature of hibernating mammals is that it can be shown conclusively that the nervous system is still active. The senses are still working, even when in the deepest sleep of hibernation, to the extent that a hedgehog, for example, will raise its spines in response to the click of the shutter release of a camera. Electrical brain rhythms can also be detected with electrodes fixed on the skull. There is a different metabolism in the nerve cells to that found in corresponding nerve cells of non-hibernating mammals.

The Origins of Hibernation. Charles Kayser has suggested that non-hibernating mammals living outside the tropics have become adapted to wide changes of temperature. They can protect themselves from cold by a relatively small expenditure of energy. Kayser has argued therefore that hibernating mammals are of tropical origin, are unable to regulate their heat loss as well as the non-hibernators, and have had to fall back on hibernation to survive. His arguments could explain one thing at least, and that is why a dormouse living on nuts and berries, which are plentiful in autumn and can be stored, must hibernate. The usual explanation for the evolution of the habit of hibernating is that it is a means of overcoming the scarcity of food in winter. This has always seemed an anomaly in other animals besides dormice. Kayser's view seems worth at least a second thought.

Echo-Location

Today it is common knowledge that bats find their way and hunt insects in darkness by squeaking and listening for the echo. The discovery is little more than 30 years old, however, and it took a century and a half to find it. There had been speculation on how bats accomplished this much earlier, but the first person to try to find out was the Italian, Lazaro Spallanzani, in 1793. He caught bats, put out their eyes and released them in a room criss-crossed with tapes. The bats flew among them without once touching them. The modern experimenter is more merciful. He blindfolds his bats. He is also able to connect his tapes to an electric circuit so that every time one is touched a lamp flickers.

In 1794, Charles Jurine, a French naturalist, plugged a bat's ears with wax and found it blundered into every obstacle. Spallanzani had conjectured that bats were guided by their own wing-beats, an idea favoured by the American Hiram S. Maxim when, in 1912, after the *Titanic* struck an iceberg in mid-Atlantic and sank with heavy loss of life, he tried to invent a safety device for ships based on low-frequency waves.

In 1920 H. Hartridge of Cambridge, England, watched bats flying from room to room through a door left ajar. He concluded the bats were using high-frequency sounds to guide them. Twelve years later a Dutchman, Sven Dijkgraaf, established that bats use echo-location and the analysis of it came in 1938, when Donald S. Griffin, a graduate at Harvard University, used the newly-invented piezo-electric crystal for detecting sounds above the range of human hearing. Griffin established that bats used ultrasonics, sounds with a frequency of over 30,000 cycles per second (30,000 KHz). Later, Griffin was joined by a physiologist, R. Galambos, to take the work further.

The history of this discovery is worth re-telling because it is typical of the course of scientific research leading to a breakthrough, for breakthrough it was. During the 30-odd years that had elapsed since Griffin's discovery one mammal after another has been found to use ultrasonics for one purpose or another. In bats ultrasonics are used for

The large ears of the Long-eared bat, with their complex structure indicate a highly specialized use of echo-location. This bat habitually flies among foliage using its echo-location to avoid the leaves and twigs. It will not only catch insects on the wing but also pick them off the leaves, using its mouth, a remarkably delicate operation in which a highly refined use is made of echo-location.

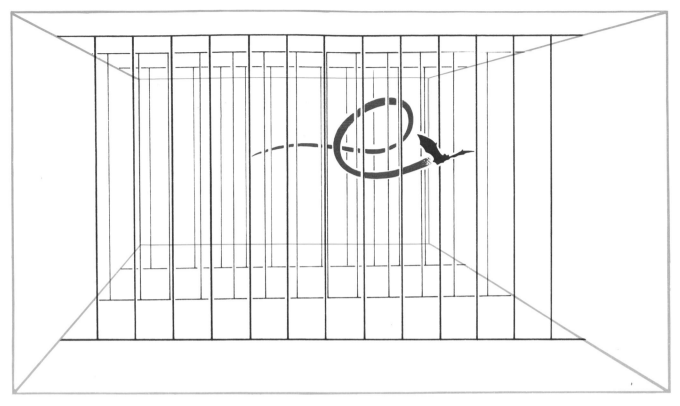

That bats can navigate using echo-location is illustrated by this experiment in which a bat successfully crosses a darkened room across which fine wires are suspended.

echo-location, the bat uttering high-frequency squeaks and picking up the echoes, so judging where and how far away an object is situated.

Echo-Location in Bats. The 800 species of bats form the order Chiroptera, which is divided into the suborders Megachiroptera and Microchiroptera, meaning literally big bats and small bats. The first suborder consists largely of Fruit bats living in the Tropics, with large eyes and good night vision. The second contains mainly small insect-eating bats with small eyes and poor vision. At least one Fruit bat (*Rousettus*) uses echo-location, but only to find its way about in caves. It makes low-frequency clicks with the tongue for this purpose. Two hundred insect-eating bats use high-frequency echo-location, the squeaks being produced by the larynx. The frequencies range from 15–150 KHz, most of them being between 20–100 KHz. Some are therefore within the sonic range, that is the range of human hearing, which is up to 30 KHz in young people but drops with increasing age. The sound frequencies used by bats have wavelengths ranging from 0·8 to 0·08 in, and the pulses vary in duration from 0·25 of a millisecond to 6·0 milliseconds.

Bats use two vocalisations, squeaks in the sonic range that express alarm, aggression and the like, and are audible especially to young people, and ultrasonic sounds known as pulses, above the range of human hearing. While at rest they emit pulses at the rate of 10 a second. When they start flying these increase to 30 a second. As the bat approaches an object the pulse rate goes up rapidly, to 60 a second, and diminishes as it passes the object. When a bat is approaching an insect the rate goes up to 200 a second. This change is necessary because the nearer the ear is to an object the more quickly will the echo come back, therefore the need for the smaller pulse.

Under test a bat can detect the presence of a wire four thousandths of an inch in diameter and can do so against considerable background noise. The echo can be heard when it is 2,000 times fainter than the noise around. The ears themselves are large, the inner ear, the real organ of hearing, is particularly sensitive and the auditory centres of the brain are relatively enormous. Even with these advantages echo-location would not be possible if the bat deafened itself every time it made a sound. Various suggestions have been made to explain why this does not happen, including the presence of a tiny muscle to close the entrance to the ear-drum as the sound is made and to open it again to receive the

117

echo. The latest is that a bat does not listen for echoes but for the beat frequency, which is audible to the bat even when pulse and frequency overlap, as happens when the rate of pulsing rises.

It has happened not infrequently that a major discovery has been made simultaneously in two different parts of the world. The most famous example is the theory of evolution by natural selection arrived at simultaneously by Charles Darwin and by E. Russel Wallace working in the East Indies. World War II broke out before Griffin and Galambos could publish their discovery. Meanwhile, in Germany, F. P. Möhres was studying the Horseshoe bats and arriving at the same conclusions. There was a difference, however. The Horseshoe bats are so named for a horseshoe-shaped flap on the face, around the nostrils. Many bats in the Tropics have these flaps on the face, known as nose-leaves. For a long time the function of the nose-leaves was not even suspected. It has, however, been investigated in the Horseshoe bat. The margins of the horseshoe can be raised and lowered and the horseshoe is used to alter the direction of the sound emissions to beam them onto objects. Also, the pulses of the Horseshoe bat

are given out in almost constant frequency, and in bursts lasting about a sixteenth of a second with only five or six to a second as against the 50–200 of the other insect-eating bats. Moreover, a Horseshoe bat emits sounds through its nose not through the open mouth as in bats lacking nose-leaves.

The probability is that the sonar apparatus of the Horseshoe bats is even more efficient than that of the bats lacking nose-leaves, if only because it gives more precise direction finding. Also, when at roost Horseshoe bats hang by the toes from a ceiling in a cave with their wings wrapped around the body, like a cloak. Other bats hang from the walls or creep into crevices. For a Horseshoe bat to hang this way it must be able to detect tiny irregularities in the ceiling in which to place its claws before actually landing. It behaves as if it can see these because it flies in, makes a quick turn to the upside-down position and takes hold with its toes, all in a minute fraction of a second. This skilful manoeuvre may owe something to the more efficient sonar.

There are still many other unanswered questions about the use of echo-location by bats. Enough is known, however, to say that a bat in flight is all the time picking up a 'sound-picture' of the world

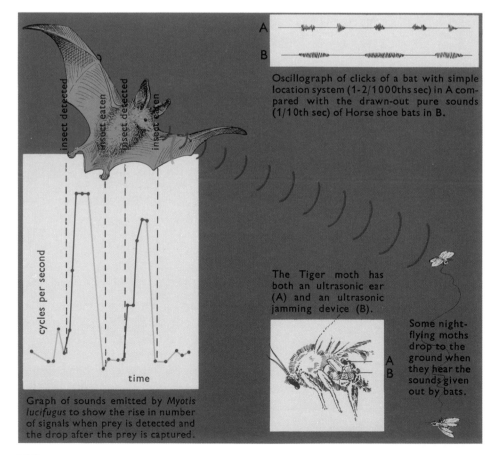

A

B

Oscillograph of clicks of a bat with simple location system (1-2/1000ths sec) in A compared with the drawn-out pure sounds (1/10th sec) of Horse shoe bats in B.

insect detected
insect eaten
insect detected
insect eaten

cycles per second

time

Graph of sounds emitted by *Myotis lucifugus* to show the rise in number of signals when prey is detected and the drop after the prey is captured.

The Tiger moth has both an ultrasonic ear (A) and an ultrasonic jamming device (B).

Some night-flying moths drop to the ground when they hear the sounds given out by bats.

A
B

The discovery of the use of echo-location by bats finally came about in the year 1940. Before many years had passed one species of bat after another had been investigated. Moreover, through the knowledge gained and the techniques devised it became clear that many other kinds of animals were using it, to navigate or find prey. In more recent years it has been established that many other animals were making use of ultrasonics to communicate with each other, or, as in some moths, to baffle their pursuers.

Dolphins can locate and catch fish using echo-location. A dome on the top of the head, known as the melon, is used in direction-finding. It is also used when the dolphin employs echo-location for detecting obstacles ahead.

around. As it cruises around its pulses, picked up on a bat detector, sound like the putt-putt of an idling petrol engine. When the bat is only an inch or so away from an insect and the pulse rate rises to 200 a second this changes to the whine of a band-saw. A bat can detect an insect the size of a pin's head a yard away or a larger insect two yards away. The Long-eared bats can hover among foliage and pick tiny insects off the leaves. Bats, such as serotines, flying in groups, can still use their echo-location in the chorus of sounds their fellows are making, and for all we know to the contrary, can recognize each other by incredibly small differences in their voices.

Echo-Location in Dolphins. While Griffin and Galambos were exploiting their discovery plans were in hand to build large concrete tanks in Florida, to be filled with sea water for keeping dolphins as in a zoo. The first oceanarium was intended as a public show but it soon became a research centre for studying the behaviour of large marine animals. Today there are seaquaria, or dolphinaria, as the later oceanaria were called, in several places in the United States, in Britain, South Africa and elsewhere. The antics of dolphins have now become familiar to millions of people.

One result of this has been the discovery that dolphins make sounds people can hear and also use ultrasonics up to 120 KHz and are sensitive to sounds up to 120 KHz. With the aid of instruments the Bottlenosed dolphin can be heard making whistling noises and rapidly repeated clicks. The clicks are used in echo-location and are emitted more or less continuously. A dolphin blindfolded by having rubber caps fitted over its eyes can swim round a tank at speed without bumping into anything. It can catch fish thrown to it without seeing them and can detect the difference between a piece of fish and a gelatin-filled capsule the same size. It will snap up the fish and ignore the capsule.

A dolphin's nostrils are on the top of the head and between these and the tip of the snout is an area known as the 'melon'. Its eardrum is connected with the exterior by a ligament instead of a tube and the two ears are not symmetrical one being placed farther forward than the other, an arrangement that

119

With the increasing number of dolphinaria open to visitors the use of echo-location by dolphins is becoming familiar to large numbers of people. The Bottle-nosed dolphin *Tursiops truncatus*, is the species mainly observed. It produces two kinds of sounds, whistling and rapidly repeated clicks, the latter being emitted more or less continuously. Early experiments quickly established that dolphins can distinguish between fishes of different sizes and a piece of fish from a similar but inedible object of the same size. They also showed the dolphins could find their way in the dark through a steel mesh curtain, avoiding gaps blocked with transparent plexiglass.

seems to be linked with direction-finding. A dolphin in an oceanarium will submit to having cups placed over its eyes but will not allow anything on the melon. While blindfolded it can detect pieces of fish in front of the melon or above it but not when they are placed under its head. Its ultrasonics, picked up on a microphone, are strongest when the melon is pointing directly at the instrument. Presumably, therefore the melon acts like the horseshoe of a bat in focusing the sounds in a beam.

Communication by Ultrasonics. If bats that fly and dolphins that swim in the sea both use ultrasonics there is no reason why many other animals may not use them, or use an echo-location system of some other kind. So the search has gone on for other examples. Several birds that roost in caves are now known to use echo-location, but with sonics, not ultrasonics. There is already a strong suspicion that several kinds of mice as well as rats, hamsters, Bank voles, dormice and other small mammals are sensitive to ultrasonics and use them either for echo-location or for communication.

The knowledge on these is somewhat scattered and fragmentary but sufficient to suggest that there is much more to be learned on the subject. Thus, marmosets make squeaks we can hear and they also use ultrasonics. We have to be content with this until somebody makes a thorough study of the vocalizations of marmosets. Young naturalists able to hear bats, when their elders cannot, also report hearing high-pitched sounds coming from shrews. There have also been a few tests made on shrews moving about in the dark and skilfully avoiding obstacles. Since the eyesight of shrews is poor this suggests the use of echo-location, probably using ultrasonics.

The European wood mouse has been thoroughly tested. The babies of this species, when between six and 10 days old, call in ultrasonics when in distress, for example when separated from the mother. She will respond by leaving the nest, even leaving the babies she is suckling, to retrieve those in distress, and a female Wood mouse carries out searching movements when a tape recording of the babies' ultrasonic calls are played to her.

White mice – albinos of the common House mouse – emit ultrasounds when touched or lightly pinched, or when they lose their balance and fall onto their sides or roll onto their backs. They do so from the age of one day but the intensity of the ultrasounds starts to decrease at 6–7 days and ceases altogether at about 14–16 days. They presumably serve as distress calls. The baby mice also react to a drop in temperature with emission of ultrasounds, at the age when they are not equipped to withstand low temperatures. The emission is feeble in very small babies, at the age when they would not leave the nest. They become strong at the age when they are likely to fall out of the nest, with the need for them to be retrieved by the mother, and decline as age increases. They cease altogether at the age when their own body-mechanism for maintaining temperature is fully effective.

Similar results have been obtained in studies with baby hamsters and baby rats.

The mother albino mouse, if deprived of her babies, also emits ultrasounds. Whether this is a call to absent babies, a sign of distress at her loss or aggression against the robber, is not yet known.

When we put these findings together and add the latest discovery in this field, that the male rat having mated signifies his unwillingness to copulate again by emitting ultrasounds, it looks as if rodents have a wide ultrasonics language. If mice, rats and hamsters use these vocalizations there is no reason to doubt a more general use among the 6,400 known species and subspecies of rodents.

Strength is given to this assertion by the fact that the use of echo-location and of ultrasonics is not confined to bats, rodents and dolphins or even to mammals. Some insects are known to use ultrasonics, and among the relatives of dolphins, the whales and porpoises, the use of ultrasonics for communication between individuals and in echo-location for depth-finding as well as hunting for

The False vampire, shown here, echo-locates by producing a few constant sounds through the mouth with which it measures speed of movement. It also emits complex sweeps through the nostrils for measuring distance.

food has been proven, or is suspected, in many species. It would be wrong to ignore, because this is a book on mammals, another previously unsuspected use because it concerns insects mainly.

Certain moths have long been known to make clicks as they fly. What is now known is that these moths have organs of hearing where the abdomen joins the thorax, each organ consisting merely of two cells. These pick up the ultrasonics of bats, which feed on the moths, warning them to take evasive action. As often as not the moths escape. There are other moths which when they click they are giving out additionally ultrasonics that confuse the bats.

Dorothy C. Dunning trained bats to catch mealworms tossed into the air by a machine. Then she played synchronously taperecordings of the clicks made by one particular moth. A bat diving for a mealworm swerved away on hearing the clicks. The moths whose clicks were being transmitted were later found to be unpalatable, so the clicks were acting in the same way that warning colours of other insects inhibit birds from eating them.

The uses of ultrasonics and of echo-location are clearly two separate phenomena which often coincide. It needs little imagination to anticipate that further discoveries in both fields in the future will open up a whole new world in animal behaviour.

◁ The Emperor tamarin, a marmoset distinguished by its long white whiskers. Marmosets are among many mammals known to use ultrasonics for communication. Others are several species of mice, hamsters, voles, and dormice, especially for communication between parent and offspring. Marmosets use ultrasonic vocalizations in addition to squeaks audible to the human ear.

Return to the Sea

Study of fossils leads us fairly firmly to the conclusion that life began in the sea and that at a later date there was an invasion of the land. Oddly enough, most of this invasion was by animals whose ancestors had colonized the rivers. The first backboned animals to live on land were beyond reasonable doubt lunged fishes. Between them and the ancestors of present-day Amphibia, the newts and salamanders, frogs and toads, there is a reasonably complete sequence among the fossils. Reptiles, and from them birds, appear to have originated from amphibians and finally shook off the shackles of having to go to water to breed. The mammals also had reptilian ancestors. Indeed, the platypus and echidna of Australasia still retain many reptilian characters in their structure, and to some extent in their behaviour.

Mammals can be said to be wholeheartedly land animals. Yet there have been at least four separate occasions when mammals have gone back to live in water. These are represented by four distinct groups: otter, seal, sea-cow and whale.

Most of the 19 species of otters live in rivers, streams and lakes. The European otter sometimes moves to the sea-shore and hunts in the sea. There is one exception, an otter wholly marine: the Sea otter of the Pacific coast of North America. Once common, but wearing a valuable fur, it was later reduced almost to extinction. Protection has enabled it to make a come-back.

A Curious Adaptation. The Sea otter lives among the kelp beds, spending much of its time floating on its back, only occasionally coming on land. The most outstanding single feature of its behaviour lies in its use of a tool. Surprisingly, this escaped the notice of scientists until some 30 years ago. When first reported it was received with scepticism but later observations have confirmed it. The Sea otter feeds on Sea urchins, mollusks, crabs and fish, diving to 120 ft (36·6 m) for them. When returning to the surface with any of the first three the Sea otter brings up with it a flat stone. Laying this on its chest, as it floats belly uppermost, it holds the Sea urchin, mollusk or crab between its forepaws and crashes it onto the stone to smash its shell. Should it dive again for food after eating it will retain and bring again to the surface the same stone.

The original scepticism arose because the use of the anvil calls for considerable manipulative skill, yet the fore-paws of the Sea otter are endowed with only stubby fingers.

Sea otters show few adaptations to life in the sea, although admittedly the rest of the subfamily Lutrinae are adapted to hunting in water. The one other big departure from the typical otter is in the permanence of its aquatic life. Otters generally, although always described as aquatic mammals, spend most of their time on land. They enter water mainly to hunt and in spite of their waterproof fur and webbed hind feet are almost at home on land as in the water.

The Origin of Seals and Sealions. Seals are another group of mammals that are aquatic yet spend a fair time on land. There are three well-defined families: the True or Earless seals (family Phocidae), the Eared seals, including sealions and Fur seals (family Otariidae) and the walrus (family Odobenidae). The three families make up the order Pinnipedia. Formerly, the Pinnipedia were considered to be a suborder of the order Carnivora (dogs, cats, etc.) but it has since been separated and elevated to an order.

It is generally assumed that seals are descended from land carnivores and there is much in their anatomy to link them with the present-day carnivores, especially in their skeletons. This is reflected in the custom of referring to their young as pups, a practice that is marred by calling the adult males bulls and the adult females cows.

In seeking the origins of living animals reliance is placed on fossil evidence. Abundant fossils of pinnipedes have been found, almost exclusively along the shores of the North Pacific. Many of these are mainly or solely teeth. Although pinnipedes have been in existence for 40 million years, since the beginning of the Oligocene period, the earliest known remains are already very similar to the skeletons of pinnipedes living today. Such differences as there are suggest that the True seals came

A group of walrus bulls hauled out on a beach. Like many ▷ aquatic mammals, walruses look ungainly on land because their bodies have become modified for life in the sea.

from otter-like ancestors and the sealions, Fur seals and walrus from bear-like ancestors. The probability is, also, that in both cases the earliest ancestors lived in fresh waters, possibly in lakes. All this is, however, highly speculative.

Nothing more can be said with certainty than that pinnipedes show strong resemblances to land carnivores and that those living today are adapted for life in water, especially in the sea. The 32 living species are distributed throughout world's seas, with two species, the Caspian seal and the Baikal seal in inland waters. The marine species tend to hug the coasts but the Northern, Pribilov or Alaskan seal spends most of the year dispersed over the North Pacific, well away from land.

Aquatic Modifications. The changes enabling pinnipedes to live most of their time in water are many. The body has become elongated and stream-lined. The tail has been lost except for a stub a few inches long. The limbs have become converted to flippers. The body is insulated by a fine fur and an underlying layer of blubber. The nostrils are valvular and can be closed under water and the external ears have been almost lost, even in the so-called Eared seals. In Eared seals the external ear has been reduced to a tiny flap. The Earless seals are, strictly speaking, mis-named because there is an even smaller ear flap but it is hidden inside the ear-hole, or more properly, in the tube leading from the surface to the ear-drum.

The teeth bear a strong resemblance to those of a typical carnivore, such as a dog or an otter, in that they include incisors and well-developed canines. There the resemblance ends for the cheek teeth (the pre-molars and molars) look alike and in many pinnipedes each tooth is tricuspid, with a stong central point called a cusp and smaller cusps in front and behind. The cusps tend to be pointed backwards and are clearly an adaptation to holding slippery prey, the fish, squid and octopus on which most pinnipedes feed.

Some of this prey is taken in shallow water but all pinnipedes can dive deep and stay below the surface for up to half-an-hour. The greatest depth recorded, for a seal caught in a net, is 600 ft (183 m). The respiratory adaptation making these deep dives possible is one of the more interesting features.

When about to dive, a seal breathes out and only a small amount of air is left in the lungs, and even that is locked by a system of valves, in the less absorptive parts of the lungs. There is, therefore, no nitrogen to cause the 'bends', which a human diver coming to the surface too quickly suffers from. While at depths of 50 ft (15 m) or more, gaseous nitrogen under pressure goes into solution in the blood. If the diver ascends too rapidly, nitrogen comes out of solution as bubbles which block the smaller vessels. This interferes with the working of the nervous system, liver and joints, causing pain, even death. The diver is said to suffer from the bends.

When a seal dives it carries with it a rich supply of oxygen. Its body contains a higher proportion of blood than in land animals. In addition to the oxygen-carrying haemoglobin of the blood there is a chemical called myoglobin in the muscles, which acts as an oxygen reservoir. The presence of myoglobin accounts for the redness of a seal's flesh. Once under water the heart-beat drops from the

Whales, dolphins and porpoises are completely aquatic and, not surprisingly, were once called fish.

Australian fur seals *Arctocephalus doriferus*. Their coat, as in most mammals, is composed of two sorts of hair, the short fine hairs or underfur and the longer and stronger guard hairs. In True seals there are not more than six underfur hairs to each guard hair. In Fur seals there are at least 20. By trapping air the underfur provides an insulating layer.

usual 55–120 beats a minute, a seal's heart-beat varying considerably, from 4–15 beats a minute. Added to this there is a mechanism in the blood-vessel system by which blood is cut off from all but the essential organs, the brain and the swimming muscles. Impure blood returning from these does not go back to the heart but is stored in large veins, to be later pumped to the heart for purification when the animal has returned to the surface to breathe a few deep breaths.

Reproduction. The pinnipedes come on land to breed. At one extreme are the Pribilov fur seals of the North Pacific, that spend seven to eight months continuously at sea, spread out over the ocean, feeding. For the breeding season they return to the group of islands off the coast of Alaska, from which they take their name. The mature bulls arrive first and as each one hauls out he occupies a small part of the beach, his territory. The cows on arrival assemble in one or other of these territories, grouping themselves around a chosen male. The young males do not breed until they are two to three years of age. They form batchelor groups inland of the occupied territories or near the edge of the sea. The cows drop their pups and mate again soon after. When the pups are old enough to feed independently the whole population takes to the sea and disperses until the next breeding season.

127

At the other extreme is the Common or Harbour seal, of both sides of the North Atlantic. In July to early September mating takes place in the water or at the water's edge. The species is probably monogamous. The pups are born in the water or on sandbanks or rocks at low tide, the pups being compelled to enter the water when a few hours old, at the latest. At other times this seal hauls out to rest at times, on exposed sandbanks.

Pinnipedes can sleep in the water, hanging vertically with the head only exposed buoyed up by inflating the neck, or below the surface and rising periodically to expose the nostrils for inhaling.

Sea cows. Yet despite their link with the land, pinnipedes are fully aquatic in their daily lives and are fully and very efficiently adapted to this end. In this, they compare more than favourably with the third group of marine mammals, known as Sea cows. They form the order Sirenia and comprise the dugong of the Indian Ocean and three species of manatee, one on the West African coasts and two on the coasts of tropical America. A fifth species, Steller's sea cow, of the Bering Sea, became

extinct, killed off by sealers for food, within a few decades of its first discovery in 1741, on the Commander Islands.

Living sea-cows are totally aquatic, even breeding in water. The dugong feeds only on Eel grass, the one marine flowering plant. Manatees will also eat seaweed and, where they enter estuaries or rivers, floating vegetation or soft herbage growing up to a foot from the water's surface along the banks.

Sea cows have lost all trace of hind limbs and the front limbs are converted to flippers. The body is rounded and barrel-shaped, and the end of the tail is flattened horizontally and spreads out to form horizontal flukes, as in whales. The skin is almost hairless, which probably explains why they flourish best in warm waters, although Steller's sea cow lived in cold waters. The nostrils are high on the snout. The external ear is lost.

Fossil evidence shows that Sea cows have been in existence much longer than the pinnipedes, probably for 50 million years. For them also there are no fossils forming a link with any land animals, although their anatomy links them with the

Bull sealion of the Galapagos islands. Sealions are Eared seals, a family to which the Fur seals belong. They differ from Fur seals in having blunter, heavier snouts, a coat with a very small number of hairs in the underfur and hind flippers with the outer digits longer than the three inner ones. Both, however, differ from True seals in being able to bring the hind flippers forward when on land. This gives them a greater agility on land and in this respect they show more obviously their descent from purely land mammals.

A young Grey seal pup shows ▷ clearly one of its most important adaptations to the aquatic mode of life. The hind limbs are modified into flippers which cannot be brought forwards underneath the body—ungainly on land, but extremely efficient in water.

elephants. They have the same kinds of teeth and the teats are pectoral. Their stomach is simple, so they are not ruminants. Their bones are dense and heavy. They live in herds and are inoffensive. They are in fact defenceless against man and have long been slaughtered for their palatable flesh. Little is known about their breeding habits.

Whales, Dolphins and Porpoises. The fourth group of marine mammals are the whales, dolphins and porpoises. Since dolphins and porpoises are merely small whales, this one word will be sufficient for most purposes in what follows. They are the most fully aquatic of all mammals. They divide naturally into two groups: the Whalebone whales forming the suborder Mysticeti and the Toothed whales or Odontoceti. The fossil record for whales is relatively meagre. As with the seals and Sea cows there is little fossil evidence to connect them with known land animals of the past but a land ancestry can be inferred from their structure.

Whales have a streamlined fish-shape. The forelimbs are converted to flippers, the supporting skeleton of which shows the undoubted humerus, radius and ulna, wrist-bones and finger-bones. The skeleton is therefore directly comparable with that of the forelimb of a typical land mammal, with the human arm, for example. The resemblance is masked to some extent by hyperphalangism, the increase in number of the bones of the fingers. Instead of the three bones of the human finger, there may be up to a dozen in the digits of whales. This makes it possible for the flipper to be increased in length and surface area without loss of rigidity.

By contrast the hindlimbs have been lost except for vestiges of the hip girdle and, in some species,

the remains of a femur. Exceptionally, a whale may be born with one or more feebly developed hind-legs. For locomotion, the tail is strengthened and thickened and ends in a pair of horizontal flukes. The tail with its flukes is moved up and down, driving the animal through the water, the flippers serving as balancers and for steering.

There are other remarkable changes in the skeleton, especially in the vertebrae and the breastbone, in the way the ends of the ribs lie free, and in the way the skull has become distorted. Relationships between fossil and living mammals are estimated by comparing their skeletons. This is difficult in the case of whales, because of these distortions. Nevertheless, it was suggested long ago, on the basis of comparative anatomy, that whales were most nearly related to the hoofed mammals. Studies in biochemistry of more recent date appear to confirm this because the proteins of hoofed mammals, particularly cloven-hoofed mammals, are markedly different from those of other mammals except the whales.

Whales, being warm-blooded, have similar problems to those of seals, in insulating the body against the low temperatures in the sea. Unlike seals, they have no hair, except occasionally a few bristles on the lips and chin, especially in unborn whales, and the skin is paper-thin. Under the skin, however, there is a layer of blubber that acts as an insulator. This may be only an inch thick in a porpoise six feet long, but it is a foot thick in the larger whales. It serves not only to keep out the cold but also to allow the escape of body heat.

This looks like a paradox until we recall that blubber is not just jelly-like fat but a fatty tissue in

The hideous face of a manatee, the Sea cow Columbus mistook for a mermaid. Sea cows, which include the dugong as well as the manatee, are fully aquatic, feeding on marine and freshwater plants. Their forelimbs are flippers. Their hindlimbs have been lost in the course of evolution, and the tail has been converted to a flattened, horizontal, spade-shaped organ. Manatees rarely leave the water and then only to rest the chin on land. If placed on land near water they can just manage to wriggle back to their natural environment.

The Small-clawed otter of south-east Asia is one of a group of land mammals that have secondarily come to depend on water for a livelihood. Otters are amphibious rather than aquatic and, while they live mainly by fresh waters, certain individuals take up residence on the coasts and freely enter the sea. They represent therefore a transition between the wholly land animal and the marine mammals.

which the cells are loaded with fat globules and bound together by connective tissue and fibrous tissue, to form a firm, compact coating. This keeps the cold out. Traversing the blubber are numerous blood vessels that can, when need arises, bring the blood to the surface to discharge excess heat to the exterior. A submerged whale cannot pant, nor does it possess sweat glands. Its tremendous muscular exertion during swimming generates heat, and the most striking way of illustrating this is to refer to what happens after a whale has been killed, and its ventilating system put out of action by death. Should there be delay in removing the blubber, the muscles become cooked meat.

Another peculiarity of the anatomy of whales, was noted long ago when whales were first being dissected. This was the presence of retia mirabilia, or wonderful networks. The usual course in the blood vascular system is for arteries running from the heart to branch repeatedly until they end in a network of very fine vessels or capillaries. The capillaries give up oxygen and food to the tissues and pick up carbon dioxide and other waste substances. They then pass insensibly into vein capillaries which join progressively to form larger and larger veins carrying the blood back to the heart to be purified.

To a slight extent in some land mammals, but to a greater degree in whales, the arterial and venous capillaries form intricate networks embedded in fat. Retia mirabilia are found more especially along the spinal cord, at the base of the brain and under the ribs. Although their function is not yet fully understood, the likelihood is that they act as reservoirs of blood that can be rapidly emptied and refilled. In whales they probably control the blood pressure during dives.

Before a whale dives it takes a number of breaths in rapid succession, then lowers its head and goes down obliquely, the last sign of this action being the appearance of its arched back at the surface. Some whales, when diving deeply, bring the tail out of water and go down more or less vertically, an action known as 'sounding'.

As with seals, whales store oxygen in the haemoglobin in the blood and in myoglobin in the muscles. They charge both fully with oxygen and recharge them on returning to the surface by another sequence of rapid breathing. That is when a whale is said to 'blow'. In drawings of the 16th and 17th centuries whales were always depicted with the head out of water and a spout of water rising from the blowhole (nostril). Later this was corrected in the written word to the 'spout' being explained as a jet of water droplets formed by condensation of water vapour in the breath on contact with the cold air.

It is only about 30 years ago that it was realized that the 'blow' or 'spout' occurs equally in the warm air of the tropics and in the cold air over polar seas. It is now suggested that it is a fine foam or emulsion of oil, water and air, blown out from the sinuous passages in the nostrils, possibly taking with it excess nitrogen.

Again as with seals, the last breath before a whale submerges empties the lungs of all but residual air, the lungs collapsing and no further oxygen entering the blood stream. While submerged the pulse rate drops from around 100 to 50 and a mechanism in the circulatory system serves to retain the blood in sinuses, or reservoirs, formed by the larger blood vessels, so that oxygenated blood goes only to the brain and such muscles as are needed for propulsion through the water. As a result whales can stay underwater without breathing for long periods, up to an hour in the Sperm whale, the deepest diver, capable of going down to 3,720 ft (1,134 m). This is a record obtained for a Sperm whale that became entangled in a telegraph cable. Whalebone whales seldom go deeper than 300 ft (91 m) but can go down to 1,500 ft (457 m). The record for a stay underwater is for a Bottlenosed whale, and is two hours.

Little light penetrates to the greater of these depths and, anyway, the eyes of whales are small. Whales have no sense of smell and they have no obvious organs for picking up vibrations in water, as is suspected for the stout vibrissae (whiskers) of seals and otters. At one time it was thought they had poor hearing. In fact, it is now known to be acute, although there are no external ears, and finding prey, certainly among Toothed whales, is by use of a sonar.

The order Cetacea includes the whales, dolphins and ▷ porpoises, which in practice are whales of different sizes. The order is divided into two suborders, the Toothed whales (Odonticeti) and the Whalebone whales (Mysticeti), the second of these being numerically the superior and showing the greater diversity. The painting shows the difference between the Baleen or Whalebone whales (1-5) and the Toothed whales (6-12), as well as some fossil whales of which little is known. Baleen plate of a Right whale (1), and of a rorqual (2). Arrangement of baleen plates in the mouth of a Right whale (3), and a Blue whale (4) with its skeleton (5). A Sperm whale (6) and its skull (7). Common dolphin (8), Killer whale (9), Common porpoise (10) showing its air and food passages (11) and its skeleton (12). The fossil whale *Zeuglodon* (13) with its skull (14) and its tooth (15).

The food of Toothed whales is mainly squid, octopus and fish, slippery prey, theoretically difficult to hold. The teeth of many of the smaller whales, including dolphins and porpoises, are small, numerous and pointed. On the face of it these seem specially adapted to holding slippery prey, yet the Beaked dolphins, which take the same food, have more or less toothless gums, often with only one pair of large teeth in the front of the mouth. Sperm whales, which feed largely on giant squids, have teeth in the lower jaw only, and these fit into sockets in the upper jaw when the mouth is closed.

Eating under water is something of which the human is capable provided he puts the food in his mouth before immersing his head, in water, then keeps his mouth shut. He must also hold his breath. Otherwise he will choke or run the risk of getting water into his lungs. Any mammal actually catching its food under water needs a special mechanism to prevent this from happening. In whales this consists of a tube leading from the windpipe into the nostrils. At the same time the entrance to the gullet is closed while the mouth is open. Between the entrance to the windpipe and the opening of the nostrils or blowhole on top of the head is a contorted tube with numerous side passages, where the nasal cavity would be in a land animal. The side passages especially contain the foam driven out when a whale blows.

The need for such an arrangement, to keep water out of the lungs, is particularly acute in the Whalebone whales. In these the teeth are replaced by baleen plates, strips of so-called whalebone hanging from the roof of the mouth. The baleen is frayed on its inner edges, and the inner face of a row of these plates looks like a tangled mass of coarse hair. Whalebone whales feed on plankton composed of small shrimp-like animals collectively known as krill. A feeding Whalebone whale opens its huge mouth and swims through a shoal of krill. Water taken into the mouth escapes through the spaces between the baleen plates, the krill being caught on the frayed inner surface. As the whale finally closes its mouth, it raises its tongue and the floor of the mouth to squeeze out surplus water.

When the ancestors of Sea otters, seals, Sea cows and whales forsook the land and its fresh waters for the sea they presumably did so to escape competition for food and living space, and in doing so were able to exploit a rich food supply. The refined adaptations they acquired are eloquent testimony to how well they succeeded, as is the size to which they grew. This last is epitomized in the Blue whale with a maximum of 100 ft (30 m) length and 120 tons weight. All was well until man developed his weapons and his ships. The near destruction of the Sea otter is paralleled by the disastrous reduction in the numbers of Sea cows, seals and whales. A moment's reflection is sufficient to make us realize that the seas and oceans of the world, at the dawn of civilization, must have been teeming with these mammals that forsook a terrestrial life for an aquatic one.

Instinct, Learning and Intelligence

Mammals are placed at the apex of the animal kingdom. This was done originally because they include the people who drafted and used the classification. Yet although the action was egocentric it was influenced by the almost unconscious knowledge that mammals generally have higher mental capabilities than all other animals. This is true despite the wide gulf between the monotremes, platypus and echidna, with a brain little more highly organized than those of birds and reptiles, and the primates, which include monkeys, apes and man.

Perhaps the one single factor that puts mammals ahead of other animals is their great ability to learn. They not only learn more, but learn more quickly, and they have better memories. They also have more insight behaviour and they have more intelligence. There are some scientists who would deny altogether the possession of intelligence by animals other than humans, but this is an extreme view.

A Few Definitions. The first difficulty encountered in discussing these aspects of animal behaviour is that of defining such qualities as instinct, insight, intelligence, even memory and learning. A reflex is easy to understand. It is a simple involuntary reaction to a stimulus. A screwed-up piece of paper is thrown at one's face and the upper eyelids drop to protect the eyeballs. The action requires no thought, nor could any thought be possible. The sight of the ball of paper travelling towards the eye causes an electro-chemical nerve impulse to travel from the eye to the spinal cord. There it is passed on to other nerve cells that cause a similar impulse to go to the eyelid muscles, which take appropriate action. A reflex action, also, cannot be controlled.

It is unnecessary here to enter into the analysis of what is commonly called instinct. This is a word that is avoided by students of animal behaviour, who prefer to break it down into its component parts, the most important of which is the innate response to stimuli.

Every animal, and mammals are no exception, lives largely by its automatic, or innate, reactions to stimuli from the environment or from within its own body. Two simple examples will illustrate this. An animal when hungry reacts by seeking food. The first step in this involves exploratory behaviour. If the animal is a carnivore it will, on seeing another animal of a prey species, react by attacking it with the aim of eating it. Nothing of this requires thought. Even when presented with a choice of food, the selection is instinctive and will depend to a large extent on the senses of smell and taste.

A Fat-tailed dwarf lemur *Cheirogaleus medius* in its arboreal habitat. Lemurs are relatives of monkeys and live a complex social life based on their ability to communicate both with sounds and visual signals.

We may go further and take the case where a hungry wolf is approaching a caribou with a view to a meal. As we have seen, if the caribou stands its ground the wolf may not attack, but will certainly do so if the caribou starts to run away. It is possible, in the first of these circumstances, to speak of the wolf as being cautious, even cowardly, and of the caribou as being determined or courageous. We might even say, when the caribou stands its ground and the wolf turns away, that the wolf seeing its prey unwilling to act as a sacrifice has 'thought better of it'.

In fact, such terms are subjective. The caribou does not 'make up its mind' to stand its ground and the wolf does not think whether it is safe or not to attack. Both react to the situation on the basis of an internal mechanism with which it was born. Either this or insight behaviour is at work, if insight behaviour were at work we could say with truth that the caribou 'realizes' that if it shows signs of being prepared to use its antlers the wolf may hesitate to attack, and that the wolf realizes that to attack would be dangerous and so looks elsewhere for a meal.

In fact, innate behaviour – or 'instinctive' behaviour – is little better than reflex action in being automatic. There is perhaps slightly more chance of its being modified. Insight behaviour has a different quality, but, before considering what it may be, it should be recalled that one of the reasons why mammals have a greater appearance of mental capacity than other animals is that their sense-organs are highly organized. Most of them lack the keen sight of birds, even the human eye is only about one-tenth as efficient as the best bird eye, but they compensate for this with better senses of smell and taste, hearing and touch. In other words, their sense-organs are collectively more effective than those of other animals.

Insight has been defined as the 'apprehension of relations' and the 'organization of perceptions'. Both definitions fail to help the untutored to comprehend. A better understanding may, perhaps, be gained by recalling an experiment carried out many years ago. A cat was put in a box and the box was automatically locked as the lid was pressed down. Inside the box were several knobs or levers one of which, when touched, released the lock so that the lid could then be pushed up.

The incarcerated cat, finding itself imprisoned, scrabbled about at random in an endeavour to free itself. Sooner or later it touched the appropriate lever. The lock was undone, the lid opened, the cat saw daylight and pushed up the lid to regain its freedom.

No doubt, if the cat were imprisoned again in the same box soon after the experience, it would either press the appropriate lever at once, or find it with fewer trials. If it did, it would have demonstrated that it was capable of insight learning, which is the solution of a problem by the sudden adaptive reorganization of experience.

That this would be likely to occur is adequately shown by the many experiments made on rats in

Left. Diagram illustrating a simple reflex tract: the sense-organ cell (1) receives a stimulus; this causes a change in the cell which is passed on through a nerve to the spinal cord (2) and at that point is switched over to a movement or motor tract (3); when the stimulus reaches the muscle (4) the muscle contracts; the reflex is therefore an automatic response to a stimulus.

Right. The fore-brain or cerebrum is amongst other things the centre for voluntary movements and sensory perception; 1. motor and 2. sensory centre for arms, legs, head, etc.; 3. primary and 4. secondary auditory centre; 5. primary and 6. secondary optical centre.

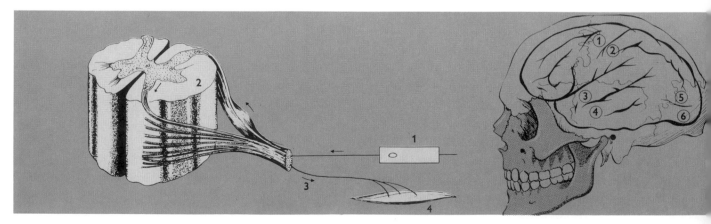

An experiment that demonstrates intelligent behaviour in a chimpanzee. On the left is a code of lights that signify the numbers 0–7 which is learnt by the chimpanzee. When it is shown a card on which six symbols are represented (A), the chimpanzee is able first to turn on three lights on a panel (B), and then to turn off one light to give the correct code for the number 6 (C). Chimpanzees that can perform such tasks show that they are able to master very elementary mathematical concepts.

cages with a choice of levers one of which, when depressed, releases a food reward. The rat soon presses the lever by sheer random movement of its paws as it tries to find a way out. Having done so it quickly learns to press the correct lever and to do so confidently. The experimenter often seeks to elaborate the experiment only to find – at least this happened on one occasion and probably occurs much more often, if the truth could be known – that the rat, even after a short experience, seems almost to anticipate his thoughts – to the secret amusement of any onlookers!

It could be argued that both the examples chosen could more justifiably be classified as instances of trial-and-error learning. This also is more marked in mammals than in other animals, and represents another important factor in the total mental equipment. Trial-and-error learning, in its truer sense, is more deliberate than in these two chosen instances. There is usually less panic and more purposive testing of the obvious lines of action. A man entering a strange room, and wishing to switch on a light, is confronted with six light switches. There is only one light bulb, so he deliberately tries one switch after another until the light comes on.

This illustration is an over-simplification to emphasize that trial-and-error learning is typically a matter of limited choices (or trials) and a more or less methodical testing of these in turn to achieve a goal. The cat locked in a box works purely at random and in a disorderly quasi-panic.

Insight can be thought of in similar terms. The sense-organs feed information to the brain. The information concerns a new situation. The information is sifted in the brain by random working of the brain cells. Sooner or later the key information, or clue, is discovered and the answer is presented.

It is generally agreed that mammals are capable of insight behaviour, however we choose to interpret this. Whether they are capable of thought, reason or intelligence is not so widely accepted. That

these are related mental phenomena is unlikely to be disputed, although the frontiers between them will be disputed for a long time yet.

Intelligence is capable of many different definitions – or of none. A professor of animal behaviour once, in a discussion, declared he would never be persuaded to define it. So much depends on the view one takes. One experienced zoologist, who had given the matter much thought, defined intelligence as the ability to recognize a problem, formulate a solution and *quickly* act upon it.

Thought Processes. One of the problems before us at this moment of writing is: what is thought? It is a commonplace experience to have a problem, to look at it this way and that, turn it over in the mind, fail to find a solution then dismiss it for the moment. What in fact we are doing is to examine all the possibilities and then submit them to the subconscious mind to work out. At a later time there suddenly comes into the mind the correct solution. The classic example is that of Archimedes in his bath, but there have been countless people since who have had a similar experience and some of them have actually shouted 'Eureka'.

What takes place in the brain, between the moment of recognizing the problem and the moment when the solution is thrown up to the surface, is a matter for speculation. In spite of all that is now known about the working of the brain the amount that remains to be known is even more formidable. It is possible, however, to speculate on the basis of observed natural behaviour linked with results of experimentation, and to compare these with what we ourselves know of our processes of thought.

Many people deliberately set an apparently insoluble problem aside 'to sleep on it'. More often than not the answer springs to mind the following morning, as we awaken, with the mind otherwise clear. Many new inventions are known to have come about in this way. It is said that the chemist who hit on the idea of linking molecules in a chemical formula with short dashes had been worrying over the problem of how to do so, was on a bus travelling through London. He saw an advertisement, then current, showing three monkeys holding each others tails. It solved his problem. If the story is not true it could conceivably be true.

Radar is said to have been invented in a similar way, by an outside stimulus, the sight of a second image on a television screen. It gave the necessary clue to a mind already working on the problem – as with Archimedes. But not all decisive clues come from external sources. Most are internal and suggest something akin to the cat locked in the box scratching around until it touches the appropriate lever.

It was said of Henry Herbert Asquith, a former Prime Minister of Great Britain, that when he had a perplexing political problem to solve he went to a particular room in the House of Commons. This had a blank wall painted white. Asquith, it was said, would take a chair and sit staring at the blank wall until the light dawned. It was a variant of the method of sleeping on a problem.

◁ Gorillas tested under laboratory conditions are found to have general intelligence of about the level of the chimpanzee. By general intelligence is meant the ability to vary actions according to experience. The gorilla is more methodical and consistent in its application to a problem; that is, it has more concentration than the mercurial chimpanzee and shows a stronger tendency to try to solve a problem for the sake of doing so than for the sake of a reward.

Rhesus monkeys *Macaca* ▷ *mulatta* taking a siesta in the trees. If vivacity were a fair test of intelligence, then these monkeys would be placed high on the list.

Some Experiments with Mammals. In English idiom we speak of 'an idea coming to us', 'the light dawned', 'the penny dropped', all illustrative of this same concept. Humans are not alone in this, as witness the experiments carried out by B. J. Beck, of the University of Chicago, in 1967, on gibbons. Dr Beck was using gibbons for experiments in problem solving. One experiment alone is of especial interest here. Two strings were attached to food. One led directly back to the gibbon's cage but the gibbons could not get the food by pulling it because the other string led away from the food, passed around a peg and only then led back to the cage where it was tied to the bars.

When the gibbons had failed to get the food because they were pulling the wrong string they gave up and went climbing around the cage. Then suddenly one would come back and pull the other string, to get the food. To all appearances the gibbons had banished the problem from their minds as being hopeless of solution. After a while the penny dropped. Dr Beck concluded this was insight behaviour. It differs only from the normal conception of insight behaviour in that there was a perceptible time lag. It was more truly Archimedian behaviour – which shows how little dividing line there is between pure insight behaviour and great thoughts leading to epic discoveries.

J. Lalande records how he spent several weeks following a huge wild boar and observing its behaviour. He describes how it would methodically block all but one of the exit holes of a rabbit burrow, with its snout. Then he would dig into the one exit left to kill and eat the rabbits. This has a touch of genius in it.

If that appears too strong an assessment for the mental powers of a hog, it should always be borne in mind that the limits of an animal's executive capacity must be considered. Thus, the results of Beck's experiments have a further interest, because gibbons had always been regarded as backward as compared with the other apes. In Beck's first experiments his gibbons had failed miserably in pulling strings, whereas chimpanzees had triumphed.

A gibbon's hands are long and slender, the fingers long and curved into hooks, and the thumb is short, although it is deeply cleft from the palm so that a greater proportion is free than in other apes or in man. It is the best kind of hand for swinging through the branches of trees but has its limitations for picking up small objects. When the string was flat on the ground the gibbon appeared stupid, when it was raised off the ground, so that it could more easily grasp it, the ape proved it was not mentally retarded.

Laboratory experimentation is not always the best method for assessing mental capacities, because the experimenter often fails in other ways to appreciate the physical limits of his subjects. It would require considerable ingenuity on the part of the experimenter to devise adequate intelligence tests for a dog. Going round watching a dog, as

Probably no other animal has been so extensively and patiently tested for intelligence than has the chimpanzee. In some of the later experiments filmed and televised, one can almost see the ape thinking in a way that challenges comparison with human beings. One of the more remarkable observations, of recent date suggests that chimpanzees can, by a means as yet undiscovered, convey abstract ideas to each other.

Polecats are carnivores that hunt their prey, usually rabbits and other small mammals, by smell. Their hunting instincts, coupled with their ability to be trained, has led to their domestication. The domestic polecat or ferret is lighter in colour and is used to hunt rabbits and vermin.

Lalande did his wild boar, is more rewarding.

A dog sniffing the ground suddenly stops and starts to dig with his forepaws. If the ground is open grassland he will soon have dug out a nest of baby fieldmice. Now watch him when he scents a nest of mice under a stump of a tree. His paws are impeded by tough roots. The finer roots yield to his paws, the larger roots he wrenches away with his teeth. He does not try these large roots with his paws but judges from the look of the roots that only his teeth and a strong pull will wrench it away.

We could argue that this is the result of earlier experience or, in other words, trial-and-error learning. Then we watch further and see how the dog, having uncovered a large root, too big for even its powerful jaws to sever, by-passes it by digging under it.

All this may sound simple, but the dog works at

high speed, throwing the earth back with paws in a continual shower, never hesitating in using its teeth on the roots or stopping to reflect on whether to by-pass the large unbreakable root. The whole operation gives an impression of intelligence which is nonetheless convincing because the results cannot be expressed in graphs or technical jargon.

This is probably why dog-owners tend to rate the intelligence of their pets more highly than does the ethologist.

An Intelligent Monkey? A most interesting observation was entered in his diary by the late Col A. Fitzgerald. He was serving with the British Army in India when one day he halted at the roadside, near a small clearing in a patch of jungle. At its centre was a pond with steep sides. A group of Rhesus monkeys appeared, led by an old male. A female slipped on the pond's edge and fell into the water. Scared, she tried to scramble out, but each time the male barked savagely at her and threatened her.

After a while the male suddenly ceased his bullying and looked sharply round. He chattered briefly to a young male in his party and the youngster climbed rapidly to the top of a tall tree. After a few moments the young monkey started chattering excitedly. The old male immediately climbed the tree, joined the youngster and, presumably having verified the report, descended again, closely followed by the young male. They joined the rest of the party, including the female, who had by now climbed out of the pond, and all moved away into the jungle, out of sight.

Three minutes later another, stronger party of Rhesus monkeys arrived at the pond, obviously determined to take possession of the territory. Fitzgerald's analysis was that the old male had become suspicious and had instructed the young male to go up the tree to survey the country around. The young male reported back, and the old male went up the tree himself to verify his report. Proof having been obtained of the presence of a stronger troop, the old male had come down and taken instant precautions for the safety of his party. Col Fitzgerald commented: 'Bully he may have been but the old male did not neglect his duty. Could a member of the human race act with more logic than this representative of the animal world?'

Col Fitzgerald was a keen naturalist. It would have pleased him to know that later studies – fifty years later, in fact – would bring strong support to his deductions, that the Rhesus monkeys could use

something approaching conceptual thought. Shortly after this W. Koehler began his celebrated tests with chimpanzees, which showed that these apes were able to use sticks to draw bananas that lay beyond the bars of the cage, within reach of their hands. They fitted sticks together to reach fruit farther away or piled boxes one on top of the other to reach fruit hanging out of reach from the ceiling of the cage.

Even these experiments were no more surprising than the observations of goats in Tunisia that are known to jump onto the backs of donkeys to browse the leaves of trees. One goat is recorded as having butted a donkey to make it stand under a tree to achieve this same end. There is a record, with photographic substantiation, in England of a donkey that habitually picked up a sack in its teeth, turned its head and beat its hindquarters presumably to drive away troublesome flies – or was it a rudimentary form of flagellation?

The point of this comparison is that even Koehler's experiments proved only that what mammals lower in the scale may do exceptionally, apes can do more frequently and with greater effect. All of it takes us no further, however, than the level of trial-and-error and insight behaviour.

Communication. From Koehler's days there has been massive research, on chimpanzees especially, and on other apes, all of which tended to narrow the gap between their performances and those of humans. During the course of this, there have been isolated instances which suggested that apes might be able to communicate by means other than vocal or gestural signs. They have been largely ignored in favour of other lines of research. Then in 1969, R. A. and B. T. Gardner, of the University of Nevada, published the results of their successful efforts to teach chimpanzees the American sign language, used by the deaf in North America.

It is tempting to investigate animals lower in the mammalian scale in order to discover if there might

The gibbon is one of the species of Tailless apes, and the ▷ most primitive. It has been little tested in the laboratory for intelligence and if it were it probably would suffer by comparison with the others. To some extent this would be because its sensory equipment has been specialized towards swinging through the trees, which it does with remarkable grace and agility. Its senses are highly alert for catching and holding branches with split-second timing. Its ability to judge distances and speeds of movement are almost phenomenal. Added to this the structure of its hands, its training and its experience, mould the hands for other work than manipulation of unusual objects, so necessary in the usual laboratory tests.

The word 'intelligence' can mean many things to many different people. One definition has been offered in the text here. Another, popular with some animal behaviourists is that intelligence is the ability to adapt behaviour to changed conditions. This is also a definition of learning, and the ability to learn must be included as intelligent behaviour whether it is shown by a worm or a man. If these assumptions are acceptable, then intelligence starts from low indeed in the animal kingdom and steadily increases as we ascend the scale. A worm can learn to negotiate a simple maze (a T-tube offering a choice of soil or an electric shock). Intelligent behaviour, through the capacity to learn, is related to brain size and complexity, reaching a peak in man.

successes

failures

conceivably be indications of similar propensities. One's thoughts go immediately to the dog that, when its owner is late in giving it its one meal of the day, goes up to the owner and very obviously moves its mouth suggestive of eating – 'smacking its lips'.

Dogs, in common with children, can use the eyes to effect. A dog will look at its owner, then roll its eyes to look at its empty plate, a clear invitation to give it food. A child will look at an indulgent aunt or grandparent and roll its eyes towards the place where chocolates are kept.

Mammals of all kinds can communicate mood and intentions by means of different calls or nuances of calls, by facial expression and bodily posture, and by scent. The number of these used, and the combinations in which they are used, increase in more or less direct proportion to the size and complexity of the brain. It has, however, always been supposed that the great difference between animals and man was that the former could not transmit information about their environment.

E. W. Menzel published, in 1971, the results of his observations on a group of eight young chimpanzees, all well known to each other. One of them was taken out of sight of the rest and shown some food, but not allowed to touch it. He was then taken back to the other seven and the whole group allowed to wander freely. When the one that had been shown the food started to go back to where it was, the others either followed or ran ahead of him to find it.

Menzel rang the changes. He repeated the experiment, this time showing one of the chimpanzees a plastic toy which induced a measure of fear. This time, after the chimpanzee had been returned to its fellows, it again went back to look, accompanied by the other seven. They now approached the hidden object cautiously and, moreover, all showed signs of fear when the object had been removed, so that there was nothing there when they reached the spot.

In other trials two chimpanzees were taken from the group. One was shown fruit, their favourite food. The other was shown vegetables, a food but less favoured. When taken back to their companions and all allowed to wander freely the whole group would go to the fruit.

The tests suggest that there had been a pooling of ideas, although Menzel had not detected any calls, gestures or any other tricks of behaviour normally used by animals for conveying information.

Grooming

A characteristic behaviour of mammals is their grooming. The most familiar example is the household cat washing itself. Grooming is the counterpart of preening by birds. On the face of it, grooming would seem to function as a means of keeping the fur clean and in good condition, just as in birds it is the main way of maintaining the plumage. If that were so it would go ill with some species in which grooming is minimal. In fact, grooming plays a part in courtship, birth of the young, social cohesion and, judging from experiments on rats, the physical and mental welfare of the individual.

Grooming and Good Health. In these experiments litters of baby rats were treated in different ways. Some litters were merely fed by hand. The baby rats were not tended by a mother. They were kept warm. They were adequately fed. In fact, they were supplied with every bodily comfort and need, but they were not handled any more than was necessary. The second group of litters were fed by the mothers, who had adequate food to supply themselves and their babies, so that each family was provided with the same comforts and necessities as was the first group. The third group, kept under exactly the same conditions as the first group were handled and petted each day.

As the rats grew they were subjected to tests for endurance and intelligence. The members of the third group emerged markedly superior. They could withstand the rigours of unusually high or unusually low temperature, they could tolerate temporary fasting or other adverse circumstances far better than the members of the other two groups. Next in order were the rats nursed by their mothers, and last were the babies that had been properly fed but deprived of natural nursing and petting.

The only important difference between the three groups of baby rats was that those of the third group had been handled and petted unduly by human fingers.

Presumably when a baby animal is fondled and stroked the surface capillaries, the fine blood vessels in the skin, are stimulated and the blood in them

A group of New Forest ponies with two foals in the background indulging in mutual grooming, nibbling and licking each other's back. Apart from any other significance or function grooming may have, it clearly is a pleasurable experience. A horse, otherwise showing a tendency to be aggressive towards a particular person, can be calmed by scratching its hair in a particular part of the body, especially on the neck or at the base of the tail.

The amount of grooming indulged in varies widely from one species to another. Wherever it occurs it tends to have a social significance, although among the hoofed animals, of which the giraffe is one, grooming between individuals, takes the form of 'necking' indulged in only by the males. This action characteristic of the species, appears to have a strong sexual significance, possibly leading to the coming together of the sexes.

flows more freely. This freer flow is communicated to the rest of the blood system and all the tissues receive more oxygen and food. Continued daily this can have a high therapeutic value, inducing a feeling of well-being and radiant health. At least, that is what the results of a daily friction bath would suggest.

The protagonists of Nature Cure recommend use of the daily friction bath, which consists of brushing the whole surface of the body with a soft brush, one that is comfortable when applied to the back of the hand. The results are clearly demonstrable to anyone taking the trouble to test the method. They include a general *joie de vivre*, clearer sight, greater energy, and other desirable effects. Surface massage

by hand can have similar effects. The stimulation of the blood flow is fundamental to all the so-called unorthodox cures, such as breathing exercises, hot and cold baths, cold water treatments, local or general massage and the like.

If these things are true, it is easy to see the damage done in childhood to the individual deprived of mother love, with its handling, fondling and petting.

If the effects of fondling and grooming can have such marked effects on physical well-being, they must also influence the brain and its performance. Even without specialized knowledge of its structure and physiology, it is obvious that in ordinary day-to-day living the brain is not used to its full capacity. There is a saying that 'fear lends wings to the feet'. Translated, this means that in moments of danger anyone will run faster than normally, or even faster than they would have thought themselves capable. This can be expanded to the general statement, that in a crisis any individual, human or animal, can excel in feats of physical strength or endurance, or, so far as the brain is concerned, in intellectual feats.

Given a clearer, more healthy brain, the 'crisis effects' become more or less permanent, in the sense that a greater use of the brain is made continuously in everyday living.

Licking the New-Born. Speculation along these lines can be endless, especially so far as humans are concerned. To see the practical effects among animals we can best return first to the question of superficial massage. For this we can watch a female mammal about to give birth.

As the moment of parturition approaches, the female has the impulse to lick. This is best exemplified by a pregnant female domestic cat who had driven one of her subadult offspring away from her by an aggressive display. A little later, as the moment for giving birth drew near, the subadult passed near her. She proceeded to lick its head and neck, in what can only be described as a bout of effusive affection. Then, as her first kitten began to descend the birth canal, the female did what most female mammals do at that moment, she started to lick her vulva. As the head of the first kitten

emerged, she licked it. This ruptured the birth membranes. The licking continued, so removing the birth membranes, freeing the kitten, but before this the licking, applied to the region of the emerging kitten's nostrils had stimulated its breathing. Parturition is a busy time for the mammal giving birth to several young. She must then lick the kitten all over, removing the birth fluids, drying it, as we say; and for each kitten the whole series of lickings must be repeated. The drying is most important, for if liquid evaporates from a surface it lowers the temperature of that surface. Maternal licking is an important factor in keeping the newborn offspring warm.

Another important function of maternal licking comes into play when the mammalian mother licks the anus of the newborn, stimulating the evacuation of the bowel. This empties the baby's intestine of waste material accumulated during the time the foetus has been receiving food material via the maternal blood, while in the womb. With altricial young, that is, mammalian babies born helpless and blind, and nursed for a period in a 'nest', the licking of the anus continues throughout the time the young are living in the nest. The intestinal content of the infant is swallowed by the mother, thus obviating fouling of the nest.

In some altricial mammals a nest of dried grass or leaves may be made to receive the newborn babies, or it may be merely a layer of hair the mother tears

A Coke's hartebeest, or kongoni, of East Africa, nibbling at an itch on its rump. How far it is justifiable to include this as a form of grooming is hard to say. The evidence is that grooming can perform a wide range of functions, including that of self-gratification.

In many species of mammals the grooming is mainly confined to self-grooming, as when a cow rubs her flanks against a post, or when a horse or a donkey rolls in the dust. Chinchillas roll in dry sand, which is unusual behaviour for a rodent. Chinchillas in captivity which are inclined to bite when handled can be rendered placid by brushing their fur with a soft brush.

from her front with her teeth. The stripping of the hair has the further advantage that it exposes the maternal teats, making it easier for the blind, helpless babies to find them.

The reverse of altricial young are the precocial young, such as calves, foals and fawns. These are the offspring of herbivores, hoofed animals feeding on grass or leaves, that must move from pasture to pasture. The young ones must be able not only to follow the mother but also keep up with her at an early age. They must be able to stand on their own feet within an hour of birth, and to be able to run as fast as the adults at an early age.

On the whole, members of species producing precocial young are less sharp-witted than those producing altricial young. Whether this is the result of less licking as babies is a matter for debate. At least it can be said they receive less, nor do they use mutual licking or mutual grooming, as it is

called, to the same degree. This whole phenomenon is only now being explored and at this stage of our knowledge it is possible to speak only in generalities, with many gaps in our knowledge.

Grooming. Grooming, is a continuation of the licking that starts with the expulsion of the foetus from the maternal birth canal. Later in infancy the babies begin to return the maternal licking. The members of a litter may occasionally lick each other, but the impulse to use the tongue becomes stronger as the animal grows up. It is a sequence easily watched in a pet puppy growing into a dog or a kitten growing into a cat. These two provide a strong contrast.

It is usual to speak of a cat washing itself but dogs have to be given a bath. A cat licks its fur more or less over the whole body. A dog licks its front paws and genitals (in the male) only. All mammals have their specific pattern of grooming. In some species

Rodents are commonly given to self-grooming, licking their fur, meticulously cleaning the tail and whiskers. Some rodents groom themselves most carefully after waking in the morning. They do so in such a way as to recall the human visit to the bathroom after a night's sleep. Grooming can also be a form of displacement activity, as in this Wood mouse embarrassed by the sight of a camera.

One White-throated capuchin grooming another. It used to be thought that grooming between monkeys was a matter of one animal relieving another of its skin vermin. Modern studies have revealed a deep social function. In monkeys it is largely restricted to dominant males or to females with infants. Grooming is solicited, also, and for each species of monkey or ape there is a particular posture indicating this. A chimpanzee stands with its head bowed, a baboon smacks its lips. In others the method of soliciting is more subtle but equally unmistakable.

it is almost or quite non-existent. And between these and the apparent fastidiousness of a cat's toilet are a great variety of intermediate behaviour patterns. It is the same with scratching, which is another form of grooming. There is a rough ratio between licking and scratching: the more licking done the less scratching, and vice versa.

There are side-effects also. It has been shown, for example, that a rabbit washing its ear with its paws moistened with the tongue takes into the mouth a natural oil which, irradiated by the sun, contains vitamin B. It is essential to the health of a rabbit to have this vitamin. A young rabbit deprived of sunlight, or deprived of the oil by having its ears washed with ether, develops rickets.

Bears are particularly given to scratching themselves and to rubbing their backs and flanks against rocks and tree trunks. Dogs and others rolling on the ground are showing the equivalent of these

actions. It is often said of the European badger that it is a very clean animal and this is usually contrasted with the fox, which is condemned as a filthy animal. The comparison is unfair and based largely on the badger digging a latrine in which to deposit its faeces, whereas a fox, like others of the dog family, deposits its droppings at random, making no attempt to bury them. So far as bodily cleanliness is concerned, badgers are no cleaner than foxes. Anyone having kept a tame fox will have noticed how it will brush itself against bushes and other foliage. It will take a few steps one way, pressing its fur against the foliage, then turn round and repeat the process for the other flank. The habit is shared with domestic dogs. It is a form of self-grooming. Certainly a healthy fox always has clean, sleek fur.

Grooming as a Means of Communication. Grooming, scratching and self-grooming keep the fur clean, but as shown by the rabbit, these are not solely a matter of toilet. It is not impossible that one effect is therapeutic, although this has not been fully investigated. Another effect is the founding and maintaining of social bonds. The mother becomes more firmly attached to her offspring, the offspring to the parent. Grooming or mutual grooming can be a feature of courtship and in sociable mammals they are a means of keeping the members of a group or pack together and of ensuring harmony between them.

A dog licks its master's hand, or his ears and face. Even a sheep will lick the hand of its shepherd. And such mutual licking or grooming is made on those inaccessible places such as the neck, behind the ears, under the chin, down the throat, and, in animals with a less supple spine, at the root of the tail. A wild ass in a zoo, aggressive towards one particular person, was converted to a more friendly attitude when that person was able to scratch its hindquarters, at the root of the tail. It would thereafter, when it saw this person approaching it, draw near and present its hindquarters to him to be scratched.

Much of what has been written so far in this chapter is now accepted as commonplace by students of animal behaviour. A great deal of it may have been learned pragmatically by those whose business it has been to handle animals. Yet the collation of it into anything approaching exact or documented knowledge is of fairly recent date although the opportunities for studying these things have been with us a long time because most aspects of grooming behaviour are displayed by domesticated animals and pets.

Man and His Pets

When a human being takes an animal for a pet he removes it from a natural to an unnatural environment. As a result the animal's behaviour changes. The changes are most marked when the pet is a mammal. The basic 'instincts' remain but tend to be in abeyance, although ready to emerge under suitable circumstances. There are, however, more important changes, conspicuous in their total effect but difficult to define or to analyse.

The first animal to be domesticated, and also made a pet, seems to have been the dog. The domestication occurred, according to the findings of bones in prehistoric middens, some 10,000 years or more ago. Some authorities put the time at nearer 15,000 years. This was long enough ago for the origins of the domestic dog still to be uncertain. There are several theories: that the ancestor was the

wolf, that the dog is derived from the wolf with the admixture of jackal, and that the dog arose from a wild species of dog that has since become extinct. When the evidence, such as it is, is sifted, the first of these theories seems the more acceptable, especially because of what has been learned of the behaviour of the wolf during the last decade or so.

It has long been known, from ordinary household observation, that yawning is infectious with dogs. It is possible to make a dog yawn by yawning at it. The act of yawning has a different function in different types of animals. In fishes it is a prelude to activity following a resting period. In humans it has this function in the early morning but at the end of the day it indicates fatigue, and in between it usually means boredom. It is infectious among humans also. Somebody once had a picture of a

It is generally accepted that the cat was first domesticated in Ancient Egypt, or, if not, that domestication reached a high pitch there. This may be no more than an illusion from the number of antiquities associated with that civilization. If the domestication of the cat occurred in Egypt the most likely progenitor would be the Bush cat *Felis libyca*, of Africa.

yawning baby hanging on her wall. She took mischievous and secret delight in watching her visitors continually yawning as they sat opposite the picture. Wolves yawn at each other. They hold yawning 'choruses' before setting off on a hunt. These are highly infectious and ensure that all members of the party are equally aroused to action and in a mood to co-operate.

Another chorus used by wolves is the howling chorus. The members of a hunting party sit or stand around, point their muzzles to the sky and howl. This is believed to create a bond between them and to synchronize their mood. Jackals and coyotes also use the community howling. L. Crisler found it was possible to make friends with wolves by joining in their community howling. So can be answered a question of long-standing: why a dog will sit and howl, with its muzzle pointed to the sky, when its owner plays a musical instrument.

These two features of behaviour give strong support to the idea that the dog is a domesticated wolf. It is not possible to do more than speculate on how or why the dog should have been domesticated in the first place. There is a similar ignornace on the uses to which it was first put, whether kept for food as dogs still are in some parts of the world, as an adjunct to hunting, as a guard dog or merely as a pet. All roles and others besides have been fulfilled by the domestic dog at various times and in various places.

The Importance of Scent to the Dog. One thing that must have been noticed, but which was markedly more noticeable as towns grew in size and clean towns became the ambition of local citizens, was the dog's habit of urinating repeatedly in the streets. Any post, wall, tree or lamp-post provided a target for this habit, which became the object of many jokes. Clearly this was not merely a matter of physical comfort since a dog taken on a lead rations its output so as to spray suitable points along the route both on the way out and on the return journey, keeping some in reserve until the last post near home.

When taken for a walk along the same route the dog will spray the same posts but if it finds, by sniffing, the signs that another dog has used a post it will over-spray it.

To the human this is an unpleasant habit. To a dog it is a vital part of its way of living. These scent marks, for that is what they are, speak volumes. In the wild they would mark the boundaries of the dog's home range. They would tell a bitch the route

a dog had taken and, although the bitch, and this is true of females of other species, uses scent marks differently from the male, these keep the dog informed of her presence and of the progress of her oestrous cycle. The persistence with which dogs on meeting first sniff noses in greeting, then sniff each other's hindquarters to serve for later recognition, is all part of an essential ritual.

With the exception of the primates, including lemurs, monkeys, apes and man, which are eyesight animals, and the marine mammals such as some whales that lack a sense of smell, the sense of smell is dominant. The world around is a pattern of smells as detailed in its odours as are its visual details to a primate. Each animal marks its world with its body scent but, more especially, with its dung, its urine and with secretions from its special scent glands. The scent glands are located in different parts of the body according to the type of animal. A he-goat's aroma may be well-nigh intolerable to the human nostrils but not to a goat, to which it conveys a significant message. Deer have special scent glands on the face or between the hoofs.

By translating the urine marks of domestic dogs to the wider sphere of the countryside at large, it is possible to visualize the landscape marked with trails and boundaries of the mammals living in it. They mark the land in plots, they cross and criss-cross. To the 'smell animals' they convert the countryside into a survey map. The freshness or staleness of the scent marks act as a diary or calendar which, with the memory of individual odours, tell the smell animals of the nearness or otherwise of friend or foe, how long since it passed that way, and other details we cannot perceive with our eyes.

Today, everyone is familiar with the idea that animals hold and defend territories or home ranges. It is little more than half a century ago that this concept was fully worked out; and it came from the study of birds. The domestic dog has been offering mankind similar information for study for 10 millenia. Because the human nostrils are so deficient, and because man so often fails to see the things that are, as we say ironically, 'right under his nose', he failed to learn the lesson. It was only by extrapolating from what had been learned of the behaviour of birds that the territorial habits of

Animals as pets are of fairly recent origin. The first mam- ▷
mals to be domesticated were those that are important
sources of food or, as in the case of this husky, animals that
can be usefully employed by man.

mammals became clear. Even then, the methods used by the smell mammals were first elucidated from the study of wild mammals after which, to everyone's surprise, it was realized that domestic dogs marking lamp-posts with their urine were demarcating the bounds of their home range.

The Origin of the Domestic Cat. The domestic cat was first allowed into man's dwelling houses little more than 5,000 years ago, and the horse was subdued at about the same time. It is not fully appropriate to speak of the cat having been domesticated because it has retained much of its independence. Its ancestry is a little more certain than that of the dog, although there is still a slight doubt whether it was the Bush cat or Cafer cat of Africa. The doubt can, however, be ignored for all practical purposes. The place of its origin is usually supposed to be Egypt, largely no doubt because the Ancient Egyptians deified it, mummified it in large numbers and portrayed it so extensively in their arts. The Bush cat is like the European wild cat, including the subspecies known as the Scottish wild cat, and all three bear a close resemblance to the common household tabby. They are probably all members of the same species.

With the cat having adopted man as a provider of food and shelter for only half the time that the domestic dog has been under man's yoke, there has been correspondingly less time for it to offer us information on animal behaviour. In addition, the cat is nocturnal, indolent and sleepy by day. Even its courtship is known more by the ear than by sight so that much of it is still only partially revealed to us.

The cat has posed problems to students of animal behaviour. Thus, it was long ago noticed that it responds more to the high-pitched feminine voice than to the lower masculine voice. The reason has been fairly recently discovered, and is that a cat is more sensitive to sounds in a range higher than is heard by the human ear. This, presumably, enables it to hear the high-pitched squeaks of mice, its natural prey. The purring of a cat has prompted many enquiries. It seems now that this is a juvenile trait which persists into the adult. Although it appears to come from the stomach it originates with the vocal cords, the rest

The ancestry of the domestic cat is in doubt. In this instance, the difficulty is that so many small wild cats such as Pallas' cat *Felis manul* from appearance alone, could be likely candidates for selection as the household cat's ancestor.

The mystery surrounding the original ancestry of the domestic cat is increased by the remarkable similarity between the Bush cat *Felis libyca*, the European wild cat *F. silvestris* and the Scottish wild cat *F. silvestris grampia*. These have long been accepted as separate forms, but zoologists are now coming to the view that all three represent one and the same species. If that is so, then domestication could have occurred at any point over a wide range of territory.

of the body providing resonance. However it is a phenomenon which so far has defied satisfactory explanation.

Cats, like dogs, can express their moods by the poise and the movement of the tail, even the pose of the body, as well as by facial expressions. Both, and the cat particularly, have a wider range of meaningful vocalizations than they are normally credited with. Much of this was unknown until the last decade or two, but the 'language' they represent is gradually being worked out, and future generations of pet keepers will be better able to understand what these and other pets are trying to convey to them. That, at least, will be one dividend from the increasing study of mammalian behaviour.

Communication Between Pets and their Owners. Dogs especially, are able to respond to their owner's words. When a dog-owner says, 'He (or she) understands every word I say', there is a tendency to underrate the truth of it. This is not to

say that dogs can tell the meaning of words. What they do is to interpret the intentions, even wishes, of their owners which the words uttered represent.

Human beings rely on the use of words to communicate with each other. Those who live in an advanced society also enjoy a high degree of security. These twin factors tend to put the use of the senses in abeyance for minute to minute living. Some dulling of the senses almost certainly occurs in domesticated animals and in those living closer to man, in his secure household. Nevertheless, they are still dependent on interpreting minute details of the environment.

Some idea of what this means can be gained from hearing those people recounting their experiences who had had to live in a world in which speech is kept to a minimum, such as inmates of prisons. They tell of a quickening of the senses so that every slight movement on the part of their fellows, every faint odour and the slightest sounds all have their

155

The vexed question of the ancestry of the domestic dog will probably never be settled. Most people accept the wolf as the ancestor but there are leading experts in the field who maintain that the Golden jackal has contributed to the lineage of the dog.

message. They become acutely aware of other people's moods and intentions.

Probably to an even greater degree are the senses of animals quickened. An experienced dog or cat, even if its senses are partially dulled by domestication and the security that goes with it, shows clear signs of responding immediately to every tiny movement, especially of the person to whom they are most attached. They become particularly responsive to those movements conveniently called intention movements.

When somebody sitting in a chair decides to stand up he first adjusts the balance of the body so as to facilitate the workings of the muscles that will translate him from the first position to the second. Outwardly there is little to show that this is taking place. These adjustments are followed by definite movements to bring him onto his feet. It may be that, having made the initial adjustments he changes his mind and decides to remain seated. His intention to do something has, however, been betrayed.

Every action is preceded by intention movements.

Changes in mood can be reflected in them but more strongly also in posture of the body, expression on the face, inflexion of the voice and gesture. How much richer is the communication between two people when standing face to face is shown by the multitudinous misunderstandings that can be conveyed in letters, and by the difficulty of sorting out a misunderstanding over the telephone.

When these ideas are applied to animals lacking speech the importance of this wealth of minutiae is easy to realize. A dog that has lived close to a person seems not only to appreciate what is required of it but to anticipate that person's wishes. Words for it are superfluous, or at best only a means of reinforcing what it has already gleaned.

Another factor must be added to this, which can be exemplified by the child brought up in an affluent household. It absorbs knowledge and skills as if soaking them up from its environment. Even an adult constantly with another adult of superior learning or intelligence increases in intellectual

156

stature. Laboratory animals, even those of small cerebral capacity, do the most remarkable things as a result of training. A laboratory rat trained to pull a ladder, by means of a cord, to reach food on an upper shelf, seems to use something approaching reasoning to do so.

Are these increased abilities all due to contact with man? Are they capacities latent in the mammalian brain, even in the lower members of the order? Or is it that animals in the wild also possess abilities, and use them in day-to-day living, undetected by man's limited observation of the world around him?

In 1950, Lt-Col J. H. Williams published an account of an experiment he carried out with his pet dog. It was in India. He sent his manservant out with the dog. Williams and the manservant synchronized watches. The manservant's orders were to watch closely what the dog did at precisely 11.00 hours. At precisely that hour Williams mentally commanded the dog to come home. It arrived home a few minutes later. When the manservant returned he reported that up to 11.00 hours the dog had been playing about in the grass. At precisely that hour it suddenly stopped playing, stood alert for a moment, then turned and headed for home.

Other people reading this tried the experiment for themselves with similar results. Moreover, many people have told anecdotes about their dogs which seem to indicate their pets' ability to know, understand or in some unexplained way to learn of their owners' intentions, commands or wishes. Do these experiences indicate a form of knowledge or communication hinted at by some writers but rejected by most?

Experiments with dolphins in a dolphinarium have shown that instructions given to dolphins in one section have been passed on to dolphins in another section, even although the two sets of dolphins are separated by a thick, high wall. The instructions are carried out by the second set of dolphins in a manner that appears almost uncanny. The comment by the scientists carrying out the experiments is simply that so far they are unable to explain how it is done.

Index

160